THE BIT ON DIGITAL COINS: A PRACTICAL GUIDE TO UNDERSTANDING CRYPTOCURRENCY & BLOCKCHAIN

DAWDU M. AMANTANAH

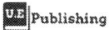

Disclaimer

Copyright 2020 © Dawdu M. Amantanah

All Rights Reserved

No part of this book can be transmitted or reproduced in any form, including print, electronic, photocopying, scanning, mechanical, or recording, without prior written permission from the author.

While the author has made every effort to ensure that statistics and information presented in this audiobook are as accurate as possible, any implications, direct, derived, or perceived, should be used only at the reader's discretion. This book is presented solely for educational purposes and is not financial or fiduciary advice. The author cannot be held responsible for any personal or commercial damage arising from communication, application, or misinterpretation of the information presented herein.

"At its core, Bitcoin is a smart currency, designed by very forward-thinking engineers. It eliminates the need for banks, gets rid of credit card fees, currency exchange fees, money transfer fees, and reduces the need for lawyers in transitions... all good things."

—**Peter Diamandis**, Founder and Chairman of the X Prize Foundation

PREFACE: WHAT CAN GO DIGITAL WILL GO DIGITAL

People who read about technology and digital transformation witness a lot of hype around blockchain and digital currencies. Although these solutions are famous for their investment opportunities, their practical value is a continual evolution underneath several layers of complex yet powerful technologies.

After stripping away the hype, it's easier to see how beneficial blockchain is, not only for exchanging digital currency securely but also for keeping records and storing data reliably and transparently. As our social and financial systems become increasingly subject to surveillance, blockchain is the only solution that practically offers freedom from it.

The potential for change is endless for blockchain. It can and will create a massive ripple in the coming years and revolutionize several industries across the board. Although it is now a bit early to say, with the power blockchain technology has, it can influence the world in the same manner as big data and the internet.

This book will give our audience a basic introduction to the blockchain where users and investors can understand basic concepts. The main goal is to help you learn practical ways to safely buy & sell digital currencies and cryptographic tokens and where to get started.

You will learn the best crypto investment and trading secrets while at the same time, handle risks related to theft & scam. Overall, this book will give people new to the industry an excellent introduction to blockchain technology and cryptocurrencies.

CHAPTER ONE

THE BUILDING BLOCK OF DIGITAL LEDGERS: HOW TO GET STARTED WITH BLOCKCHAIN

If you have been following cryptocurrencies, investing, or banking over the past decade, you may have heard about blockchain as the record-keeping technology that created the Bitcoin network.

For someone outside banking or technology, getting your head around jargon like "distributed databases," "public ledgers," and "decentralized authentication" is problematic. Fortunately, the breakdown of blockchain is a lot easier to understand than most people think.

What is Blockchain?

At its very core, blockchain is simply a computer file for storing data and record keeping. In more technical terms, blockchain follows a ledger of open and closed distributed databases. Blockchain's purpose involves storing data by duplicating, dividing, and holding it across multiple computers simultaneously.

. . .

What makes blockchain different from other means to store your data (i.e., cloud or hardware storage devices) is that the system is entirely decentralized and anonymous at the same time. In other words, no single person, company, government, or corporation has complete control over the entire blockchain. This system is radically different from centralized databases and record-keeping systems, like banks that can control, administer, trace, and identify all transactions made across their network. Due to its anonymous nature, you can carry out transactions without the threat of being drafted and traced.

Why Do We Call It Blockchain?

Blockchain technology gets its name from the way it stores information in individual blocks of data. All these blocks are connected to a previous block, forming a union chain, thus creating a "blockchain." Each block of data contains a digital piece of information comprising of three main properties.

Date, Time, Money Spent

Blocks store information like date, time, and dollar amount of your most recent transaction, just like your Amazon transaction, for example, when you order an item on their website. These records also store information about the people participating in these transactions.

Anonymous

. . .

In the blockchain, blocks don't record your name or your vendor's real name. All parties participate anonymously, and records are stored using a unique "digital signature," something similar to a username.

Identifiable through Hash Codes

Every block has personally identifiable information that distinguishes it from other blocks. Like us, these blocks have names in the form of unique codes called a "hash." Hashes are cryptographic codes that developers create with the help of individual algorithms.

These codes are essential to maintain an accurate and consistent record of transactions and ensure transparency. For instance, consider that you are purchasing something from Amazon, but you can't resist repurchasing the same item as the item is in transit.

When this happens, your new transaction details will be identical to your previous purchase, but we can tell both these blocks apart because of their unique codes.

Although we are considering a block equivalent to the record of a single purchase from Amazon, it's slightly different in reality. For instance, one block on the Bitcoin blockchain can store as much as 1 MB of data. The number of transactions a block can keep depends a lot on the size of the transactions. Therefore, a single block can store thousands of transactions.

How Does Blockchain Work?

Whenever a block stores new data, it gets updated on the blockchain. However, because blockchain has multiple blocks strung together, you must do four things to add a new block:

You Must Perform a Transaction

. . .

To get this point across, let us consider your purchase at Amazon again. You can't control your inner shopaholic and purchase something you don't need.

Once you buy that item, your transaction will remain recorded by your username. However, since a block can house thousands of transactions, the blockchain will package your transaction with records of other user's transactions made at that time.

The Network must verify Blockchain all transactions

After every purchase, you must verify the transaction you made. Usually, in public record-keeping, there's always someone responsible for verifying new records and data entries. It happens everywhere from Wikipedia, the Securities Exchange Commission, even at your local library.

However, with blockchain, a network of computers has to carry out this job. For instance, once you purchase an item on Amazon, this network of computers will rush to verify whether the transaction happened the way you claimed. These computers will match the transaction's time, the dollar amount spent, involved participants, and all other details of the purchase. (We'll explain how this happens shortly)

That Block Must Be Given a Hash

The block will not become a permanent part of the blockchain unless all its transactions are verified. As soon as the computers verify all transactions, that block will earn a unique, identifying hash code. Just as your degree verifies that you passed all your program exams, the hash code verifies that all these transactions are valid and authentic.

Additionally, the block stores the hash code of the most recent block added to the blockchain (the block added before our new block).

After the hashing is done, our new block becomes a permanent part of the blockchain.

Interestingly enough, every verified block is publicly visible to everyone –even you. If you want to, you can look at Bitcoin's blockchain that contains information about Bitcoin's entire transaction data. You can see important information such as where ("Height"), when ("Time"), and by whom ("Relayed By") a particular block was joined to the blockchain.

Blockchain Privacy?

While a blockchain's contents are public, you can also connect your computer as a *node* to receive updates about new blocks added to the blockchain. A node is simply a part of the network of computers that verify transactions in a blockchain.

How Secure is Blockchain?

Your computer will automatically save a complete copy of the blockchain and update you about new transactions. Every node (computer) connected to the blockchain network stores its copy of the entire blockchain. There are thousands, and even millions of copies of one blockchain, if it's as large as Bitcoin. Initially, it seems unnecessary to store so many identical copies of the entire blockchain, but these copies are essential to prevent records manipulation.

For instance, if you have ten Bitcoins stored under your username and a hacker wants to steal it, it will be challenging for him or her to do that. Even if he hacks a machine and puts all your Bitcoins into his account, this won't transfer your Bitcoins.

The hacker will have to change every copy of the record that says that you own those ten Bitcoins. So, he must hack millions of computers simultaneously to steal your Bitcoin from you. Because of this sparse and secure record keeping, blockchain for security is a "distributed" ledger that can be trusted.

That said, not every blockchain makes identifying information in transactions public. For instance, you'll notice that the Bitcoin blockchain doesn't give us access to identifiable information about the users making transactions.

While blockchain ensures that you stay anonymous on the network, your personal information remains limited to your username or digital signature. If we don't know who adds blocks to the blockchain, how can we completely trust the blockchain or the network of computers storing that information? Leading us to our next question:

Blockchain technology implements several measures to maximize security and trust in the system. First, the blockchain stores all new blocks, chronologically and linearly. In other words, a new block is added to the "end" of the blockchain, which is the last added block before a crypto owner can link a new one.

After being verified, every block has a position on the blockchain, which we refer to as its "height." If we glance at Bitcoin's blockchain, we'll see that it stores the height of every blockchain. As of August 2020, Bitcoin has surpassed a height of *643500*.

Once we add a block to the end of the blockchain, it's complicated to modify or remove that block's contents. Every block stores its hash and the hash of the previous block creating converting our information into a string of letters and numbers. Therefore, if we try to edit the information of any block, its hash code will change.

This feature helps a lot in improving the system's security. For instance, if a cybercriminal tries to edit your transaction to make you pay for the purchase twice, as soon as he tries to edit the dollar amount in the transaction, the block's hash code will change.

However, since the next block still contains the old hash, the hacker needs to repeatedly update its value until it reaches the entire chain's last block.

Therefore, to edit a single block needs to recalculate all hashes stored in all of the subsequent blocks, which takes an incredible amount of computing power. Cracking the block equation is extremely difficult to edit and almost impossible to delete it once added to the blockchain.

In terms of trust in keeping data safe, blockchain networks follow several tests for computers that want to become nodes on the network. A computer must qualify all these tests for any node that wishes to gain the ability to join the network and add blocks to the chain.

These tests are formerly known as "consensus models," where users must "prove" themselves before becoming a part of the blockchain network. Bitcoin uses a system known as the "proof of work", where computers must solve complex puzzles to "prove" that they have done some "work."

As soon as a computer solves one puzzle, they are eligible to add blocks to the blockchain. The process of adding blocks to the blockchain is popularly known as "mining," and being able to do it is extremely difficult.

As of January 2020, the odds of solving these puzzles on the Bitcoin network were nearly one in 15.5 trillion. Given that Bitcoin is an exclusive network, these odds are not at all surprising. Because these puzzles or mathematical problems are incredibly complex, Bitcoin miners have to spend a significant amount of computational power and energy to solve them.

Due to its complexity, the proof of work system makes any hacker attack useless. If a cybercriminal wants to coordinate an attack on the blockchain successfully, he must control more than 50% of the blockchain's total computing power. Currently, China has the largest mining operations in the world regarding Bitcoin. Millions of Bitcoins are stored offline, so executing a 51% attack on its blockchain successfully via mining or a computer is not worth the effort and nearly impossible to obtain.

This mechanism is necessary for a blockchain network where all participants are anonymous. Although generating proof of work is very expensive, it is essential to build trust between parties that have no reason to trust each other.

CHAPTER TWO

Introduction to Cryptocurrencies

As it seems, blockchain is a complex system where thousands and even millions of people use their time, effort, resources to ensure that blockchain runs uninterrupted. However, to keep the system running, it's essential to incentivize people who invest in the system. Otherwise, why else would someone dedicate a significant portion of resources to create blocks?

Digital Tokens vs. Cryptocurrency

Several miners compete to solve a mathematical puzzle. We know whoever solves the puzzle first gets to build the block on the blockchain. The concept of *cryptocurrency* began as a reward for miners who dedicate their resources to solve the puzzle as quickly as possible. Mined blocks have become valued highly by all blockchain participants. We take digital coins equivalent to electronic money and use it to fund various activities on the blockchain. These specialized coins and tokens are the only valuable item in the blockchain; thus, its record use as a unit of value.

Although some people use "Coins/Cryptocurrency" and "*digital tokens*" interchangeably, they are different terms used in other contexts. Both refer to digital currencies, but the primary difference is in how we record their transaction. We build tokens and host them on existing blockchains, whereas coins are assets, which have their own dedicated, standalone blockchain.

Bitcoin (BTC), Ether (ETH), Litecoin (LTC), and Ontology (ONT) are classic examples of coins. One thing common in all these cryptocurrencies is they all have their blockchain network for recording transactions.

For tokens, the Ethereum platform is one of the most well-known examples of token-based general-purpose blockchain technology. Other token platforms are NEO, Omni, Stellar, and EOS.

Another main difference between these digital currencies would be how there used by members of the blockchain. People use cryptocurrencies like Bitcoin to pay for services and goods. They like to trade it for other currencies and even store it as an investment.

Although we mainly use cryptocurrencies for monetary transactions, coins have other use cases too. For instance, someone can use an Ether coin to make transactions on the Ethereum network. Developers and investors receive tokens from the Ethereum network, but the Ethereum network is the train tracks needed to send a token and fund transaction verification on the Ethereum blockchain.

While coins do have additional functionality, they are mostly just a means of payment. Digital tokens, on the other hand, have a more comprehensive functionality. Digital token use is mainly in general-purpose blockchain, where people and businesses are creating "*smart contracts*" (a form of binding agreement) in blockchain software.

Thanks to the **Ethereum smart contract**, anyone can now use tokens to reward miners for processing transactions. The smart contract concept has helped people create their forked tokens for personal and business usage.

All they have to do is to issue their tokens after making

an **ICO** (Initial Coin Offer). You can use these tokens to anonymously pay for services or hire anonymous miners to record your transactions on the blockchain. However, tokens don't have personal blockchains, and we must record them on the coin ledger based on the token connected to it.

To summarize, only cryptocurrencies can save their transactions on their blockchains, whereas tokens rely mostly on other blockchains—the reason why the demand for coins of established blockchain networks is so much in-demand.

At this point, many people wonder, "Why must we create a separate currency when we already have traditional currencies to do that online?" Well, the short answer is that traditional money doesn't work on blockchain. Most importantly, traditional paper cash doesn't fit a currency's role to exchange in a transparent, anonymous, and secure manner.

Let us dig deeper on why is that the case:

Cryptocurrency vs. Traditional Currency

Cryptocurrencies are similar to digital cash, designed to be cheaper, quicker, and more reliable than government fiat-backed money. In traditional currency, you rely on the government and banks to create your money and the means by how it is stored, sent and received.

Everyone involved in this process takes incurs a fee to process your transactions. On the other hand, cryptocurrencies allow you to transact directly with other users and store your money securely by yourself. As you no longer need a go-between to send money instantly, crypto transactions are usually faster and cost less than they do with regular banks.

The cure to minimize manipulation and fraud, blockchain networks allow us to simultaneously record and verify our transactions, as well as the transactions of other people. Everything is traded and recorded on a public ledger. Additionally, everyone involved can tally records to keep them authentic, secure, permanent, and transparent.

The blockchain public ledger technology powers the legitimacy of cryptocurrencies. This system has the potential to revolutionize transactions and disrupt traditional payment systems. For instance, the banks store records for investment options such as bonds, stocks, and other financial assets digitally, and they demand a trusted third party to verify transactions.

What's unique about this system is that it doesn't require trust between any party, so some people call it the trustless system. Because all the records are public, you no longer need to trust your money with a bank.

You don't even have to put faith in the person you are trading cryptocurrency with, as nothing happens without your approval, and there is no way to counterfeit a false agreement.

All the while, you can watch your money being issued, sent, received, verified, and recorded by all the members of the blockchain network (computers that verify the transaction). You don't even have to become a miner yourself to buy, secure, and trade a cryptocurrency.

What Determines the Value of Cryptocurrencies

The value of each cryptocurrency depends on many factors. Besides the declared value of a cryptocurrency, people who invest in the currency also rely on a perceived "inherent value." These factors include how satisfactory the technology and network are themselves, how reliable and secure the currency's cryptographic code is, and how extensive that decentralized network will operate. As a precaution, before investing in any asset, you must understand that the cryptocurrency market is volatile. Crypto gains its legitimacy from stakeholders, merchants, consumers, tech developers, regulators, and financial institutions.

Consumers and Merchants

. . .

Cryptocurrencies offer faster and cheaper peer-to-peer payment compared to the traditional money exchange and service businesses. Cryptocurrencies increasingly gain acceptance as payment options among the consumers and the market.

Interestingly, although the price acceptance and stability will improve once we use digital currencies to purchase goods and services, most digital currency owners still use it for trading. A cryptocurrency's volatility and the speculative nature of its cost currently provides greater incentive for trading instead of using it for consumer purposes.

Only 6% of respondents to PwC's 2015 Consumer Cryptocurrency Survey say they are either "very" or "extremely" familiar with cryptocurrencies. We anticipate that familiarity will increase as consumers begin to access innovative offerings and services not otherwise available through traditional payment systems. The upside of businesses and merchant's perspective, cryptocurrencies offer low transaction fees and lower volatility risk resulting from nearly instantaneous settlement.

Investors

Investors are usually confident regarding the opportunities for cryptography and cryptocurrencies. The "inherent value" of the underlying technology, discussed above, gives these investors good reason to be optimistic. As a result, only recently have some of the more established cryptocurrency companies attracted institutional investors and Wall Street attention.

Tech Developers

Many talented tech developers have devoted their efforts to cryptocurrency mining, while others have focused on more entrepreneurial pursuits such as developing exchanges, wallet services, and alternative cryptocurrencies. In our view, the cryptocurrency market has only started to attract talent with the depth, breadth, and market focus needed to take the industry to the next level. However, for the market to gain mainstream acceptance, consumers and corporations will need to see cryptocurrency as a user-friendly solution to their everyday transactions. Also, the industry will need to develop cybersecurity technology and protocols.

Regulators

Government attitudes worldwide are inconsistent when it comes to the classification, treatment, and legality of cryptocurrency. Regulations are also evolving at different paces in different regions.

Financial Institutions

Traditionally, banks have connected those with money to those who need it. This middleman position diluted in recent years, and disintermediation in the banking sector has evolved rapidly. Resulting in the rise of Internet banking, increased consumer usage of alternative payment methods like Amazon gift cards, Apple Pay, and Google Wallet, and advances in mobile payments.

Recently, cryptocurrencies such as Bitcoin have demonstrated their value effectively in the market. The Bitcoin blockchain currently boasts 18.5 million Bitcoins in circulation. With all the use cases for cryptocurrencies, it offers the most value in terms of investment.

CHAPTER THREE

BITCOIN – HOW ONE MAN'S BRILLIANT IDEA SHOWED US THE FUTURE OF BANKING

In 2008, a mysterious yet brilliant individual or group introduced Bitcoin to the world. No one knows the identity of the originator under the alias Satoshi Nakamoto. Satoshi Nakamoto generated the Bitcoin white paper via Bitcoin.org and created and deployed Bitcoin's original reference implementation. In 2009, Bitcoin was the first cryptocurrency to come into being using the cryptographic triple entry accounting system, and it built the foundation for blockchain for everyone to follow.

Bitcoin introduced cryptocurrencies to the world; people followed its example. Bitcoin is a new form of money that is two different things at the same time. It is both a currency and a transparent payment network. The idea of a purely peer-to-peer version of electronic cash that allowed online payments to be sent directly from one party to another without going through a financial institution was revolutionary, and most initial cryptocurrencies were inspired by Bitcoin. Gaining a full understanding of why Bitcoin became so popular, it is important to understand the backdrop behind it.

It isn't a coincidence that Bitcoin appeared as a store of wealth after the 2007-08 global recession hit the entire world. The financial uncertainty alone was worrisome for people across the globe, but as accusations on big banks piled in, many of them became distrustful of banks.

Wall Street accused banks of misusing borrowers money, rigging the system, and charging exceptionally high fees. The pioneers of Bitcoin wanted to become empowered, prevent double spending, abolish interest fees, eliminate middlemen, and make all transactions transparent.

At the same time, they wanted to minimize corruption; cut down transactional fees, and create a self-sustaining organic network value. This is why they created a decentralized system, where people could control and save it, exchange, and use it without depending on banks.

Cryptocurrencies offer faster and cheaper peer-to-peer payment compared to the traditional money exchange and service businesses. You don't have to submit any personal details, just public keys. Cryptocurrencies increasingly gain acceptance as payment options among the consumers and the market.

Since its inception, Bitcoin has come a long way to become the currency of the future. Companies, services, and businesses across the world, from private hospitals in Europe to large jewelry chains in North America, accept Bitcoin as payment. Even billion-dollar tech giants like Microsoft, Dell, PayPal, and Expedia, haven't shied away from investing in Bitcoin.

Many people still don't know what Bitcoin is. Its speculative, volatile nature has been shown negatively by the media and politicians since it rose from the ashes after the economy burned during The Great Recession, but that is all changing.

Bitcoin has a capped amount of 21 million coins. 18.5 million of those coins have already been mined into circulation, leaving just about two and half million Bitcoin left to mine over the coming years and decades. The naysayers will claim 21 million is not enough for the billions of people to share, own, or hold, but Bitcoin gives a solution.

Bitcoin has staying power and is virtually instantaneous. Being highly divisible, you do not need to own a full Bitcoin; you can own a fractional share of it called Satoshi's. There are 100 million Satoshi's per Bitcoin, which is plenty to go around even if you purchase small amounts.

A SATOSHI IS THE SMALLEST UNIT OF BITCOIN, EQUIVALENT TO 0.00000001	
1 SATOSHI =	0.00000001 BTC
10 SATOSHI =	0.00000010 BTC
100 SATOSHI =	0.00000100 BTC
1000 SATOSHI =	0.00001000 BTC
10K SATOSHI =	0.00010000 BTC
100K SATOSHI =	0.00100000 BTC
1 MILLION SATOSHI =	0.01000000 BTC
10 MILLION SATOSHI =	0.10000000 BTC
100 MILLION SATOSHI =	1.00000000 BTC

In the future, when the Bitcoin surge becomes more common place, the global population will only dream of owning one Bitcoin due to scarcity. The world's growing population will turn towards Bitcoin as a wealth store, with alleged estimates as high as 8%. There are claims one Bitcoin could be worth $228.6 million in a few decades.[1]

If you have a fixed supply of an asset and demand increases, the only thing that can happen is to produce more of the asset, or the price has to go up. The complete opposite of inflation based fiat currency, Bitcoin is deflationary and can only increase in value since banks cannot create more.

Several websites have emerged that publish content related to Bitcoin, detailing its price actions and recent developments. Many platforms on the web act as forums to discuss cryptocurrency and make Bitcoin trading more efficient. Publications like Bitcoin magazine have even created Application Programming Interface (APIs) to display the running price index and exchange rate.

Bitcoin has now 'forked' into more than one type of coin. In other words, the blockchain has split into a different blockchain, and forking signifies a splitting of the chain on which Bitcoin operates, diverging into two or more blockchains with different rules than the existing blockchain as the two would now have different versions of Bitcoin technology.

This happens in all blockchain cryptocurrencies and happens when the number of transactions is overwhelmingly high. Forking allows us to change the number of transactions permitted in a single block, helping us store more transactions and scale transactions easily.

However, not all stakeholders agree on changing the rules for creating blocks. All stakeholders democratically choose whether a change in rules will change or not. If a significant number of stakeholders agree on changes and others don't, stakeholders negotiate and diverge the blockchain into two different pathways.

One pathway follows the old rules, and the other follows the new ones. This creates two blocks with the same height, something that is never possible in normal circumstances. You will have the choice to consider which option is better for you, so you're not trapped at the whim of others. This also happens when developers try to update the security rules for the blockchain.

The forking process already happened for Bitcoin over a hundred times. There have been several successful forks in Bitcoin thus far, with the following forks being some of the major ones::

- Bitcoin Cash (BCH)
- Bitcoin Gold (BTG)
- Bitcoin Diamond (BCD)
- Super Bitcoin (SBTC)
- Bitcoin Atom (BCA)
- Bitcoin Core (BTX)
- Bitcoin God (GOD)
- Bitcoin Private (BTCP or ZCL)

Bitcoin does have a unique set of problems, one mainly being speed. Regardless, for most people, Bitcoin is the most reliable channel to exchange money. Therefore, it's highly likely that Bitcoin will emerge as one of the most important technologies to shape the future of banking.

Key Highlights for Bitcoin

- October 31, 2008: A man with the alias Satoshi Nakamoto anonymously published the Bitcoin white paper.
- January 3, 2009: The Genesis Block or block number (the first block in blockchain) one is created.
- January 12, 2009: The first real-world Bitcoin transaction occurs, where Laszlo Hanyecz buys two pizzas for 10,000 BTC.
- December 16, 2009: Bitcoin's Version 0.2 is released.
- November 6, 2010: The entire market cap value exceeds $1 million USD.
- October 2011: Bitcoin forks for the first time to create Litecoin.
- June 3, 2012: Block 181919 was created with 1322 transactions. It is the largest block to-date.
- June 2012: Coinbase launches.
- September 27, 2012: Bitcoin Foundation is formed.
- December 4, 2013: Price reaches a high of $1,079.
- December 7, 2013: Price falls to around $760.
- February 7, 2014: The hack in Mt. Gox– one of the worst hacks in Bitcoin, trigger a drop in price.
- June 2015: Bit License gets established, revamping the security standards for Bitcoin and making significant cryptocurrency regulations.
- August 1, 2017: Bitcoin forks again to form Bitcoin Cash.
- August 23, 2017: SegWit gets activated.
- September 2017: China bans BTC trading.
- December 2017: The Chicago Mercantile Exchange (CME) and CBOE Global Markets (CBOE) launch the first Bitcoin futures contracts
- December 2017: Bitcoin price reaches its all-time high
- January 2018: Price drops after the 2018 cryptocurrency market crash.
- September 2018: Cryptocurrency value collapses from January 2018's peak
- November 15, 2018: Bitcoin's prices stabilize after falling from $100 billion for the first time since October 2017.

- October 31, 2018: 10-year anniversary of Bitcoin
- May 11, 2020: 3rd Bitcoin halving[2]
- Present – Bitcoin continues to be a high risk, high-profit investment

Transactional Properties of Bitcoin

Secure

Bitcoin funds are locked in a public key cryptography system. Only the owner of the private key can send cryptocurrency. Strong cryptographic technology and the magic of big numbers make it impossible to break this scheme. A Bitcoin address is more secure than Fort Knox.

Permission-less

You don't have to ask anybody to use cryptocurrency. It's just a software that everybody can download for free. After you installed it, you can receive and send Bitcoins or other cryptocurrencies. No one can prevent you from transacting with your own money. There is no gatekeeper, middleman or banker.

Immutable

After confirmation, a transaction can't be reversed. By nobody. And nobody means nobody. Not you, not your banks, not the president of the US, not Satoshi, not your miner. Nobody. If you send money, you send it. Period. No one can help you, if you sent your funds to a scammer or if a hacker stole them from your computer. There is no safety net.

Anonymous

Neither transactions or accounts are connected to real-world identities. You receive Bitcoins on so-called addresses, which are randomly seeming chains of around 30 characters. While it is usually possible to analyze the transaction flow, it is not necessarily possible to connect the real-world identity of users with those addresses.

Fast and Global

The transaction is propagated nearly instantly in the network and is confirmed in a couple of minutes. Since they happen in a global network of computers, they are completely indifferent to your physical location. It doesn't matter if I send Bitcoin to my neighbor or to someone on the other side of the world.

Bitcoin The Currency Of The Future?

The future has long been touted to be upon us, but we only ever accept this until the next big thing arrives. Cryptocurrency is undoubtedly a step into the future that eliminates from these institutions mishandling of public documents and bank records, the dishonesty, lack of trust of employees, and the lack of reliability.

. . .

The fact that the currency or payment system exists in the public domain with the protection of all the nodes connected shows just how much Bitcoin can be a game-changer in terms of securities, risk management, and data management. The Bitcoin system is ahead of its time for sure with the use of an auto-updating mechanism that makes it almost impossible to breach the public ledger.

Only those with the keys needed will access their data as opposed to the archaic bank methods. The use of Bitcoin to manage the internet of things shows its application can go beyond the technical world and into the everyday world.

Economists can further see Bitcoin as a currency as it possesses three properties that define a currency: hard to earn, the supply is limited, and easy to verify. Let's expound on this a bit deeper; Bitcoin has finite scarcity, making it more scarce as time goes on, something money has never achieved before. The more people who adopt Bitcoin, the fixed supply halved, making it harder to obtain, and that scarcity fundamentally drives price demand for Bitcoin.

The currency of the future must provide for various currencies and be universally accepted to bridge gaps between people of race, gender, and transactional capacity. Various banking authorities warn Bitcoin users that they are not protected by chargebacks or refund rights, but that has not diminished its popularity.

Role of Bitcoin in International Banking

Blockchain is one of the fascinating concepts of Bitcoin, and this is bound to affect an industry that has often been known for its age-old operations and legacy record keeping.

Bitcoin is fantastic for banks as it can reflect the records of a client, the creditworthiness of a client, and prevent double transactions.

It is a safe manner of moving around money and other securities without inconveniencing banks or authorities that are subject to time-consuming protocol and scrutiny.

The Bitcoin auto-refresh system is a major attraction for banks as it prevents double transactions and, with the private and public key authentication system, can prevent identity theft and reduce the computational requirements for banks.

The downside to all this is the assumption that Bitcoin and its maker never envisaged; bad morals. Miners often offer their computational power to create blocks, but what happens in the case of a selfish miner?

What about the use of a Bitcoin to entice two offers for the same commodity in a race attack?

People can often decide to alter the ledger by omitting entries, so what happens in the case of history modification?

These are among the major worries of banks when implementing these new technologies. Their vulnerability is usually high, but especially so, in the presence of dignitaries and high-risk clients. Thus the reluctance to change their tried and trusted methods, but Bitcoin may have the best chance of lasting solutions to the international banking domain.

Proof of work is the concept of any transaction being accepted into the network meeting the difficulty target where miners find a nonce to hash with the block. One miner has to verify that a certain amount of computational effort has been expended and energy to solve the block.

Bitcoin's proof of work is completed when miners compete against each other to complete transactions on the network and get rewarded in Bitcoin. Wallets are steady credential storage devices that hold the keys to the network for any given transaction and are key in ensuring the fungibility of Bitcoin.

Zero coin and Dark Wallet protect privacy and address the issue of transactions bearing marks of identity, a key concern of the international banking community that can be hosting offshore companies.

The sensitivity of these companies and their said directors and assets are of paramount importance, and the Bitcoin technology and blockchain functions must comply with this before it is embraced.

Will Big Banks Accept Bitcoin Or Its Blockchain?

The sheer fact that it reduces the cost and complexity of the bank's transactions shows versatility is limited to the blockchain.

This shows the difference between the trust levels of human interaction as opposed to the telemetry available for the blockchain. This can be a significant flaw in its operations, but the affected companies shall step forward and continue the rapid emergence, rather than the identities breached.

The presence of modern options like multiple access points, stabilized and detailed bank accounts, or even mobile banking should surprise any economist; what was the hold up before banks acknowledge the perks of Bitcoin and blockchain.

Big Data is also one of the newer technologies being accepted by the banks, showing a slow yet sure progression. The best-case scenario involves complete acceptance of what Bitcoin and its accompanying technology have to offer.

The fact that resources can be saved by getting integrated circuits specific to the application which outperform standard CPUs at a fraction of the power consumed makes this a worthy risk field—some of the banks seem to be complicit or pro-Bitcoin include.

What More Can We Expect from Bitcoin?

. . .

Developments like those at PayPal and other intermediaries like MasterCard are great news for the crypto space; blockchain and crypto organizations should continue to collaborate with incumbents.

Much has been made written and spoken about the recent foray by PayPal and others into the cryptocurrency space, but that is just the tip of the iceberg in terms of blockchain and crypto asset development.

Adding 300 million customers who will be able to transact with crypto, and do so on a peer-to-peer basis, is good news for the ecosystem at large. Such developments might seem antithetical to the original idea and concept of Bitcoin, but are critical to the continued expansion and development of crypto assets.

Bitcoin was an ideal, and while that ideal has not exactly worked out as planned, there are several developments that continue to accelerate blockchain and crypto adoption. Intermediaries and third parties might have been the players that crypto was designed to disrupt, but in order to actually get cryptocurrencies to be used as legitimate fiat alternatives there does seem to be a need for these intermediaries to be involved. Stable coins and CBDC's are simply symptoms of a much broader trend toward more semi-centralized and centralized blockchain and crypto asset options.

Let's take a look at just why the blockchain and crypto space needs, and will benefit from, the involvement of third parties and intermediaries.

Functionality

Crypto was designed to be a legitimate alternative to current fiat currencies, but in order for that to actually come to fruition these options need to be as convenient and as simple to use as existing options. Linking in third parties, payment processors, banking institutions, or some other sort of institution will help make this possible. In the aftermath of the Bitcoin price bubble of 2017, multiple peer-to-peer services and platforms have emerged, so in order to achieve mainstream adoption, crypto options will need to be as customer friendly as these current tools.

Venmo, Zelle, and Cash App should be leveraged to help make conducting crypto transactions, including being able to reverse or edit crypto transactions. Mistakes happen, and consumers need the confidence to ensure they can undo these mistakes.

Regulatory Clarity

The rise of stable coins, asset-backed coins, or other forms of central bank digital currencies might strike many as the antithesis of the idea of cryptocurrency. As appealing a slogan as that might be, that is only a partial view of the situation; to get cryptocurrencies and blockchain at large to go mainstream, there is going to be a need for increased regulatory clarity. By working with established payment processors and financial institutions, all of whom already have experience dealing with the numerous compliance and regulatory rules, the pace at which crypto regulations are resolved will only accelerate.

Depending on which counter party, individual, or institution is asked, the bringing together of blockchain and crypto organizations with established incumbents can be seen as good or bad news. Bucketing these developments into one single category, however, represents an incomplete view of the marketplace as well as one that will hamper the continued growth of the ecosystem.

Stability

Engaging third parties will help encourage broader usage of cryptocurrencies as fiat alternatives and not just as investment options. Price stability has long been an issue for truly decentralized cryptocurrencies, but by involving some of the major payment processors, price volatility will hopefully become less of an issue. By working with, as opposed to against incumbent financial institutions and third parties, crypto assets will gain greater stability and greater utilization.

Prices for specific crypto can be higher or lower than others, but having the backing and infrastructure of well-known payment processors can help reduce some of the more stomach-churning price volatility.

Cryptocurrencies and digital money at large are the future of money; that much is clear, but in order for cryptocurrencies to fully generate and create the promised benefits, it is increasingly clear that incumbent institutions must be brought into the conversation. Regulatory experience, price stability, and the increased clarity with which crypto can be treated and reported are simply a few of the benefits that can be derived from such an arrangement.

Blockchain and crypto are the future, and in order for that future to play out as promised, there needs to be a partnership between crypto and incumbent financial institutions. This collaboration should be celebrated, encouraged, and accelerated if crypto ever hopes to achieve the widespread utilization that much of this promise is based on.

---

1. Bitcoin is considered Digital Gold because like gold it's divisibility is limited. Gold has been recognized as a store of value for centuries but not for poor people. Bitcoin will change how people harness wealth solely through a smartphone and internet connection giving it the possibility to surpass gold.
2. What is halving?

    Every 210,000 blocks mined, or about every four years, the reward given to Bitcoin miners for processing transactions is cut in half. This cuts in half the rate at which new Bitcoin is released into circulation. This is Bitcoin's way of using a synthetic form of controlled supply that halves every four years until all Bitcoin is released and is in circulation.

CHAPTER FOUR

What Problems Does the Blockchain Solve?

The blockchain technology behind Bitcoin cryptocurrency has captured the imagination of industries outside the tech sector, from the manufacturing supply chain to banking and finance. Blockchain algorithms provide us a reliable and secure way of sharing mission-critical information between organizations.

Blockchain has the potential to solve the major business problems that have been troubling owners from the onset of the decade. The scope of blockchain applications extends to several industries. Here are some of the main ways it can help us:

Reducing Data Loss

Have you or someone you know had a computer hard disk fail? If you didn't maintain a recent backup, the result is catastrophic data loss. Corporate data centers commonly use so-called RAID drives (Redundant Array of Inexpensive Disks) to address this problem. RAID drives divide and copy portions of the data onto each of its several hard disks in a very innovative way. If one disk should fail, enough information is replicated on each surviving disk to recreate the original data lost on the failed disk.

Blockchain technology provides an analogous solution. However, instead of relying on multiple hard drives within a central RAID array, the data itself is *distributed* among the users, with no central, official repository. That means that if a user experiences a significant hardware or connectivity failure, the system can proceed without him or her. How does this work exactly? Each time a transaction records (a stepwise function known in blockchain terminology as a "block"), all the underlying data (e.g., the total of ledger transaction records from the beginning) replicates onto each user's machine via a peer-to-peer network. Suppose a user misses an update due to a hardware or connectivity problem. In that case, they download a fresh copy of all the "blocks" in the historical chain of transactions to get synchronized again.

Preventing Data Corruption

Single points of failure, such as a hard disk crashing, are usually easy to detect if the errors are reproducible. On the other hand, intermittent issues (particularly ones involving multiple hardware and software systems) can be devilishly difficult to diagnose.

This poses a problem for system designers. What do you do when there is an error? The answer depends on the scenario. Suppose you play an audio CD, which hits a bit of missing data due to a scratch or manufacturing error. In that case, the player can often detect the mistake and interpolate the remaining data to "fudge" the missing sound. You might not even hear the difference. But with life-critical systems, such as modern jetliners that rely on sensors and computer software-based automation to move the flight control surfaces on the wing, rudder, and horizontal stabilizer, you need to build in more redundancy by adding in duplicate computers and sensors. If all their data is in agreement, everything is fine. But when there is conflicting, unreliable data (a scenario known as "the Byzantine General's Problem" in computer science), they must take measures to create a "fault-tolerant" design.

How does blockchain handle faults when confronted with data errors? At a basic level, blockchain stops in its tracks when it encounters a request that appears to be invalid. In other words, if it seems there is a mismatch in the data in the transaction ledger shared between two parties, no block update request can propagate across the network. In Bitcoin's case, these checks prevent monies from being transferred more than once (double spending).

Resolving Data Security and Creating Trust

Hardware and software errors are not the only data security issue. The blockchain has to also protect itself from willful fraud and theft.

Two blockchain design features help in this effort. The first is the immutability of the historical transaction ledger. In other words, given how blockchain is structured, it's challenging to change the numbers in old transactions. In the case of Bitcoin, this means that you can't go back in the records and add a million dollars into your account − to do so would require changing all the other transactions within the ledger, so the money was in your account and not in someone else's.

The second design feature distributes networking. Even if you could change your ledger's values, all the other users have their copy of the ledger transaction records, and the blockchain would identify this discrepancy and invalidate your account, thus preventing you from making a fraudulent transaction.

In general, it's easier to trust blockchain transactions because the ledger is public and available to all users. Contrast this to centralized systems, such as those found at a traditional bank. What happens if the bank makes an honest mistake that costs you money? As a consumer, it's more challenging to prove your case because you don't have equal access to the bank's ledger.

Fin-tech

Banking, finance, and insurance companies are working on next-generation finance technology solutions, collectively known as Fin-tech. Many are exploring cryptocurrency solutions along the lines of Bitcoin, Ethereum, and other so-called altcoin technologies. Fin-tech companies are joining forces to secure a greater understanding of the innovation of blockchain. As stated in the *Wall Street Journal*, Forty leading financial institutions have come together as part of the new R3 consortium to develop financial services applications based on blockchain's distributed ledger technology. Institutions such as the one's in the consortium maintain financial records accurately or allow insurance companies to keep a secure history of claims.

Government

Government officials worldwide are looking at blockchain technologies applications (the Federal GSA has a resource page for government employees). Among the ideas under contemplation are blockchain-based voting systems that would theoretically be more resistant to election fraud, as well as strategies to streamline and automate the collection of customs duties at borders (between Ireland and Northern Island, for example, once Britain leaves the European Union). Politicians, such as U.S. Presidential candidate Andrew Yang, have made blockchain part of their government policy campaign platform.

Our client specializes in advanced research and development for defense, intelligence, and homeland security applications.

. . .

Supply Chain Management and Manufacturing

Blockchain technology promises to reduce fraud, waste, and abuse in the supply chain. Supply chain management is a public ledger technology that could help keep trading companies "honest" while providing a more secure way to track the "chain of custody" in industries. That demand is needed more than ever, <u>a critical need to certify that items inventory are genuine and unadulterated</u>, including pharmaceutical medicines, medical devices, aerospace and aviation parts, and more. Secure, accurate tracking systems based on blockchain ledgers would make it easier to trace failures back to their source, allowing for faster corrective action quickly.

Retail Sales

Blockchain transactions not only offer a secure way to share secure data with distributors, but they also provide a fast way to share data with customers. For example, CRM and customer loyalty programs could be moved into a secure blockchain ledger, allowing transactions to be processed automatically without going through a central processing agency.

Human Resources

Companies need to collect and maintain significant amounts of data for each of their employees, including salary records, bonuses, benefits, vacation time accrued and used, insurance coverage claims, and expense reimbursements. Blockchain technology offers the potential of maintaining a secure log for each employee in a single place.

. . .

Use in Cryptocurrency

Blockchain forms the bedrock for cryptocurrencies like Bitcoin. As we explored earlier, currencies like the U.S. dollar are regulated and verified by a central authority, usually a bank or government. Under the central authority system, a user's data and currency are technically at the whim of their bank or government. If a user's bank collapses or lives in a country with an unstable government, their currency's value may be at risk. These are the worries out of which Bitcoin intended creation emerged. By spreading its operations across a network of computers, blockchain allows Bitcoin and other cryptocurrencies to operate without the need for a central authority. Risk is reduced but also eliminates many of the processing and transaction fees. It also gives those in countries with unstable currencies a more stable currency with more applications and a more comprehensive network of individuals and institutions they can do business with, both domestically and internationally (at least, this is the goal.)

Healthcare Uses

Health care providers can leverage blockchain to store their patients' medical records securely when a medical record generates into the blockchain, which provides patients with the proof and confidence that no one entity can change it. These personal health records could be encoded and stored on the blockchain with a private key so that they are only accessible by specific individuals, thereby ensuring privacy.

Property Records Use

. . .

If you have ever spent time in your local Recorder's Office, you will know that the process of recording property rights is both burdensome and inefficient. Today, a physical deed must be delivered to a government employee at the local recording office, logged into the county's central database and general index. An individual must reconcile the property's claims with the public blockchain record index in a property dispute.

This process is not just costly and time-consuming—it is also riddled with human error, where each inaccuracy makes tracking property ownership less efficient. Blockchain has the potential to eliminate the need for scanning documents and tracking down physical files in a local recording office. If property ownership is stored and verified on the blockchain, owners can trust that their deed is accurate and permanent.

Use in Smart Contracts

A smart contract is a computer code built into the blockchain to facilitate, verify, or negotiate a contract agreement. Smart agreements operate under a set of conditions that users agree upon; granted, those conditions satisfy the stipulations and the terms of the contract.

Say, for example, I'm renting you my apartment using a smart contract. I agree to give you the door code to the apartment as soon as you pay me your security deposit. Both of us would send our portion of the deal to the smart contract, which would hold onto and automatically exchange my door code for your security deposit on the rental date. If I don't supply the door code by the rental date, the smart contract refunds your security deposit. This contract eliminates the fees that typically accompany using a notary or third-party mediator.

Uses in Voting

. . .

Voting with blockchain carries the potential to eliminate election fraud and boost voter turnout, as was tested in the Nov. 2018 midterm elections in West Virginia. Each vote would be stored as a blockchain block, making it nearly impossible to tamper. The blockchain protocol would also maintain transparency in the electoral process, reducing the personnel needed to conduct an election and provide officials with instant results.

Weighing the Pros and Cons of Blockchain

Many blockchain enthusiasts on the market are glorifying blockchain technology to be the next best digital revolution. They aren't all wrong. In reality, technology does change legacy systems, gets rid of intermediaries, and introduces a new world to us. However, many do fail to properly weigh the pros and cons of blockchain before they invest.

There will always come advantages and disadvantages; blockchain is on that list as well. So, when enterprises try to figure out blockchain's full potential, they often forget about the issues.

As a result, many tend to make mistakes, lose resources, and waste time. So, to help you weigh the pros and cons of blockchain, I'll be covering them in this guide.

So, let's see what the blockchain pros and cons are! But before we jump into the blockchain technology pros and cons section, let's take a quick recap of what blockchain technology is.

Pros of Blockchain Technology

There's a lot of pros and cons of blockchain technology, but among them, we'll look into the pros section at first. Check out the pros of blockchain from below.

. . .

Disintermediation

First of all, with blockchain, you get a distributed system. It means that it gets rid of any middlemen from your system. But how is that an advantage? In reality, intermediaries tend to be the third party source that connects you to your services.

However, in the business world, from every service the middlemen offer, they get a cut. In reality, a small amount of payment might not seem like much, but it does add up when your service requires a 10-15 step process.

Other than that, there's no way of knowing whether the middlemen would be honest with their services. In reality, corruption runs deep, and in many cases, these intermediates tend to abuse enterprises and consumers for their gain.

Thus, by getting rid of them, the whole trust issue is solved.

High-Quality Data

Blockchain technology offers a superior level of data quality. In reality, it is a distributed ledger system where it stores data. But how does it provide high-quality data? Well, you need to know that low-quality data would not convert into high-quality data within a night. That's not how it happens.

Anyhow, this distributed ledger technology offers a consensus process that allows you to filter out any insufficient data with useful data. It means that no one can add any information to the ledger or even manipulate the existing ones.

So, when large swathes of information input into the blockchain, it must pass verification before getting added to the ledger, thus eliminating false data.

Moreover, it also gets rid of the issues that come from human-made errors. Mistakes are inevitable with massive data files, but there's no scope for a human-made error due to the ledger.

Thus, it significantly increases the quality of data.

Anyhow, let's move on to the next advantage in this blockchain pros and cons guide.

Durability and Security

Blockchain offers durability at its best. You could think of it as the internet where there's built-in robustness. In reality, the overall structure of the technology makes it so durable. Moreover, as it stores blocks of information around the network, it makes sure that there no single point of failure or any single entity controlling it.

This quality makes the system inherently durable. More so, as no one can alter the blocks, it remains a stable, secured platform. Other than that, it's quite efficient in fending off hacking attempts as well.

So, there's little to no possibility of overpowering this network. Anyhow, let's see the next advantage in this blockchain pros and cons guide.

High Level of Integrity

Another great advantage of blockchain is the level of integrity. Compared to any other network systems out there, blockchain offers the highest level of integrity so far. But what does it mean? In reality, it means that all your data will always be the right one, and no one can alter them once it's on the ledger.

More so, the process of storing the information and consensus processes are also robust. Moreover, any user can't just make changes to the verification as he/she pleases. Thus, it would offer accurate and reliable data every single time when you transact or store any other information.

On the other hand, every blockchain's hash ID plays a massive role in maintaining this property.

Immutability and Transparency

For the next advantage in this blockchain pros and cons guide, I'll elaborate on transparency and immutability. Blockchain comes with an immutable storage system where you can't change any single form of data or let alone delete them completely.

In reality, cryptographic hashing plays a massive role in maintaining an immutable structure. As every single block will have a Hash ID, any changes to that block's data would change the ID drastically. And it's impossible to recreate the same Hash ID again.

Thus, if anyone tries to change the data, all the other users would notice it right away. Anyhow, most of the ledger system on this technology is also open for everyone to see. Even in private blockchains, there is standard ledger information that anyone can see anytime.

Longevity and Reliability

Another great advantage of the blockchain guide's pros and cons is the reliability and longevity of the data. As you already know, blockchain is immutable, transparent, and offer integrity. All of these characteristics result in the reliability and longevity of the technology.

Furthermore, as no one can change the blockchain's rules as they please, it remains intact. More so, as it can offer viable solutions for various business issues for the long term, it becomes a reliable technology.

Many enterprises are already considering altering their legacy networks with blockchain for the long term.

Simplistic Ecosystem

. . .

Many of you may find blockchain to be a complicated ecosystem; however, trust is a simplistic ecosystem. In reality, every enterprise needs to go through various stages to process or offer their consumers a solution.

What blockchain does here is to shrink down the various stages of processing into a few steps. More so, as the technology can offer almost everything online, it's much easier to maintain.

There are already many blockchain-based solutions on the market where they offer user-friendly interfaces.

Empowered Users

It's another one of the pros in this blockchain advantages and disadvantages guide. In the traditional centralized system, the users don't have that much control over their information. As a result, multiple corrupt personnel may try to misuse the information for their gain.

However, as blockchain ensures a peer-to-peer network and gives users back the control, it ensures no one can misuse them in any way. The level of security and transparency makes sure that every individual is in total control of his/her information.

Faster Transactions

It also offers faster transactions compared to traditional means. Usually, the centralized banks can take a lot of time even to process a transaction. It became more eminent when someone tries to send money overseas.

In reality, it could take up to six days to even process that transaction. So, many consumers can't rely on traditional banks' slow-paced system in times of an emergency. However, with blockchain, you can complete a transaction within a few seconds!

That's faster than any traditional means so far.

Lower Transactional Costs

Besides offering faster transactions, it also provides a lower transactional cost—obviously, nothing for free. When you use traditional methods to transact daily, you have to give them some form of fee for their services.

Even though the fee might be small, but after a multi-step process, it may become heavier in your pocket. On the other hand, blockchain only offers a lower transaction fee in exchange for a faster transaction process.

New Business Model and Value Chain

Another great benefit of blockchain in this blockchain advantages and disadvantages guide is the new business model. Blockchain comes with a unique perspective on how we should model our business in the modern world.

For example, if Bitcoin's blockchain fails, nothing will happen; if Bitcoin succeeds, it will change the business world entirely with this tech's use, new marketplaces are forming, and new opportunities follow closely behind. Furthermore, it's a better infrastructure that gets rid of a lot of issues we face nowadays. It also adds more value to your enterprise once you start to use it by increasing revenues and promoting trust.

Improved Traceability

The last advantage in the blockchain advantages and disadvantages list so far. Well, it's more for enterprises that deal with the complex nature of the supply chain or any other similar industry—tracing back what or how you managed to produce an item is crucial.

In every stage, you need to ensure that it offers the highest quality. However, if you don't know your goods' origin, you can't provide quality. However, blockchain ensures that you can trace these items from the same source to the endpoint.

Not only useful, but you can verify any asset in asset exchanges as well.

Blockchain Technology Cons

In these advantages and disadvantages of the blockchain technology guide, I'll now talk about the different blockchain technology cons. Let's see what these are.

Redundant Performance

Redundancy is the first con in the advantages and disadvantages of the blockchain list so far. In reality, the computation needs of this technology are more repetitive than centralized servers. Every time the ledger is updated, all the nodes need to update their version of the ledger.

It's because the distributed nature of the ledger system mandates that every node should have a copy of the ledger system. Thus, it needs to undergo the same process over and over again.

Complex Signature Verification Process

Another con in our advantages and disadvantages of the blockchain list is the signature verification process. Basically, for every transaction in the system, you'll need a private-public cryptographic signature verification.

It then uses the ECDSA (Elliptic Curve Digital Signature Algorithm) to ensure that the transaction happens between the correct nodes. Thus, every node needs to verify its authenticity, which can be a tricky and complicated process.

Private Keys

To transact on the network, you will need to own a private key. Even though other users can see your public key, a private key is much more crucial as it remains hidden. Furthermore, all the blockchain addresses will have a private key.

You need to keep your private key secured by any means if you don't want other people to misuse your assets. However, if you lose your private key, you will lose access to your funds on the network, as well. There's no way to recover them anymore.

Integration Concerns

It's another one of the significant blockchain technology cons. It's mainly for enterprises that run legacy networks. In reality, blockchain would replace legacy networks. However, the integration process is still not fully functional. More so, many blockchain technologies can't work along with the legacy networks.

It means that to use it properly, companies would have to get rid of their legacy networks for good completely.

Uncertain Regulations

It's a major con in our pros and cons of blockchain list. In reality, not all blockchain technologies come with the proper set of regulations on the network. Thus, many don't trust the system at all. On the other hand, the lack of regulation brings in the concept of ICO scams.

And needless to say, many have fallen victim to ICO scams as there is no regulation related to cryptocurrencies. Governmental institutions also struggle to adopt it as this sector runs entirely on rules.

Large Energy Consumption

To ensure that every transaction is valid, it needs to go through consensus processes. The consensus process requires a vast amount of effort to form every node. Not to mention, all the nodes need to communicate back and forth to ensure that a transaction is valid.

On the other hand, consensus algorithms such as proof of work require a lot of computational power, increasing overall power consumption. However, you'd be happy to know that now multiple consensus protocols consume much less energy.

No Control for Enterprises

Well, enterprises need a specific authoritative process to use blockchain technology. Unfortunately, public blockchains won't be able to offer these controlling aspects anytime soon. However, the rise of private and consortium blockchains seems to provide both the technology's control and distributed nature.

So, we should be able to a more modern approach to this issue much sooner than expected.

Privacy Concerns

. . .

Another major con in the pros and cons of the blockchain list is privacy issues. Enterprises need their privacy to maintain its brand value at all costs. They can't reveal their sensitive information to the mass people or their competitors.

And so, many enterprises aren't that keen to use blockchain for the enterprise. However, the use of private transaction options and user-specific authentication is solving this problem bit by bit.

Cultural Adoption/Disruption

Culturally blockchain will be the great disrupter shaking up many industries across the globe. Our current business models have been running on a specific architecture for quite some time now. But the invention of blockchain threatens that system.

In reality, blockchain already started to change how the system works and has disrupted many industries. As a result, multiple marketplaces have already become obsolete.

Lack of In House Capabilities

As crypto technology is a relatively new concept, there aren't many capable developers working on it. When developers working for fintech companies try to develop their enterprise blockchain solutions, finding a qualified team to handle the project becomes difficult.

However, the rise of blockchain as a Service can help you in blockchain business strategy. These services offer high-end developers and marketing teams to help you get your blockchain solution on the market.

High Cost

. . .

Yes, blockchain is much cheaper than other infrastructures. However, it can be a costly solution, as well. The cost depends on what type of feature you want to add and your needs. More so, developing a solution form of starch requires a hefty amount of money.

Not to mention replacing the legacy system would also cost a lot of money. However, you can overcome this issue if you keep your solutions at a minimum. Moreover, joining a consortium or using a BAAS can help you out as well.

According to a recent 2019 survey on 1386 global enterprises, it may surprise you what their adoption barrier can be. You can check out the percentage form below. However, you should know that the survey does not include all the enterprises in the world that are working on blockchain technology.

Thus, the "true" percentage will always vary. If you want to get more information, you can check out the Deloitte survey form here.

Blockchain is a relatively new technology right now with a long journey ahead. So, it is quite eminent that it will have cons and pros as well. I've highlighted almost all the pros and cons of blockchain in this guide.

However, you should note that blockchain has already resolved most of the issues and figuring out a newer way to minimize the problems as much as possible. So, give it a couple of years more, and you might see a perfect technology ready to disrupt the industries.

CHAPTER FIVE

How Is Cryptocurrency Taxed?

Historically, most people have always struggled to understand cryptocurrency. However, if you throw taxes into this mix, it becomes even more confusing for people. For example, Coinbase one of the largest digital currency exchanges in the world boasts 40 million users and growing. Many crypto customers on that platform may want to know does Coinbase report holdings to the IRS? In short, it depends if you meet certain taxable factors such as executing trades or receiving crypto profits through Coinbase earn, USDC Rewards or staking. This problem isn't limited to U.S. crypto traders, but people all over the world.

In the last five years, cryptocurrencies have gained significant popularity. Since the start of the year 2020, this popularity is only increasing, causing governments to pay closer attention to cryptocurrencies. Recently, the IRS (Internal Revenue Service) has released new cryptocurrency tax guidance with that, they sent several thousands of warning letters to crypto investors who didn't comply with their rules.

All this begs the question: "How does the government tax cryptocurrencies?"

Although starting anonymously, the significant share of cryptocurrency transactions is transparent today, especially in Bitcoin. In the past, lucrative currencies such as Bitcoin have gone through surges through black-market trading. Before, governments had a limited role in dealing with these surges, now they have imposed anti-money laundering requirements.

Exchanges, too, now impose these restrictions on traders to avoid penalties from regulators. Taxes have been a significant challenge for Bitcoin traders. Although federal judges, central bankers, and regulators categorize cryptocurrencies like Bitcoin differently, they all agree that the government must tax it. The trend of cryptocurrency tax is prevalent across all major countries.

Here is what it means for cryptocurrency traders:

Cryptocurrency and Taxes: What We Know

Firstly, we must remember that nothing matters until we put it into the law. Defined laws eliminate any room for speculation and make it clear what financial regulation means. At that point, no person can unilaterally alter the tax code or redefine an asset.

However, little has changed in the United States after the IRS presented its first set of laws for cryptocurrencies in 2014. IRS Notice 2014-21 issued by the IRS categorized virtual currencies as property.

The new IRS rules state anything you buy through a digital currency will be taxed as a capital gain, regardless of whether it was in your position in the long or short term. Your taxed amount will depend on how long you held that item. Let us see what that means.

Cryptocurrency as Property

. . .

To most people, cryptocurrencies look a lot like money: they're a means of exchange and have an agreed-upon value. However, the IRS neither considers cryptocurrencies as currency and nor as securities at the moment. They consider Bitcoin and other cryptocurrencies as property, but that is rapidly changing.

Cryptocurrencies, similar to other capital assets, make you liable to capital gains and losses after using or trading them. The regulators can tax profit off cryptocurrencies as a short-term gain if you have held them for a year or less before selling. In case you have owned it for more than a year, regulation can tax it as a long-term gain.

Holding a coin for profit means that the IRS will consider any loss you gain as a deductible capital loss. Some time ago, it was possible to plead ignorance to crypto tax laws. However, now that is no longer possible after the introduction of Form 1040. The new form inquires taxpayers about whether they own virtual currencies or not.

The yes/no question related to cryptocurrencies is similar to the one that was implemented years ago on offshore investment accounts and led to the nasty treatment of people who lied.

In computing a gain or loss, you use your starting point, the "basis" of an asset, tax lingo for your original purchase price (after, occasionally, some adjustments).

While cryptocurrency brokers aren't required to issue 1099 forms to clients, traders are supposed to disclose everything to the IRS or face tax evasion charges. Taxable transactions include:

• Exchanging cryptocurrency for fiat money, or "cashing out."

• Paying for goods or services, such as using Bitcoin to buy a cup of coffee for example.

• Exchanging one cryptocurrency for another cryptocurrency.

• Receiving mined or forked cryptocurrencies.

The following are not taxable events according to the IRS:

• Buying cryptocurrency with fiat money

• Donating cryptocurrency to a tax-exempt non-profit or charity

• Making a gift of cryptocurrency to a third party

• Transferring cryptocurrency between wallets such as a Trezor or Ledger Nano.

. . .

Sales Are Not the Only Form of Taxable Transaction

You have to report the disposition of a virtual coin if it is:
- sold for cash,
- traded for another crypto, or
- utilized to buy something.

But merely transferring coins, such as from a wallet to exchange or vice versa, is not a disposition. Nor do investors who buy and hold owe a tax.

The Tax Code's Wash Sale Rule Does Not Apply

The wash sale rule forbids claiming a loss on the sale of a security if you bought that security within 30 days before or after. If, for example, you buy a Tesla share at $800, sell it at $720, then buy it back quickly, the $80 loss washes.

Bitcoins and the like are not "securities." They're pieces of "property." So you can go out at a loss and then right back in without losing the right to immediately claim the loss.

Exchanges Squeal

Coin exchanges based in the U.S. file information returns on customers with a lot of trades. The 1099-K is mandatory for a customer who, in one calendar year, does at least 200 transactions with proceeds totaling at least $20,000. The 1099-K form is rather like the 1099-B that stockbrokers file, except that the latter form doesn't have the 200-trade minimum, and the K probably won't tell you what your cost basis was for a coin.

As with 1099-B forms, so with the Ks, the fact that you didn't get the form (because you didn't do a lot of trading or for any other reason) does not absolve you of the obligation to report all sales and other dispositions.

Forks Can Create Ordinary Income

When a share of stock splits in two, by and large, there's no taxable transaction. Its purchase price gets carved up and assigned to the two pieces; you declare a sale on either of those pieces only when you dispose of it. If and when you sell a portion of the stock again, you'll get the favorable capital gain treatment. Similar examples would be if one share of Exxon Mobil split into one share of Exxon and one share of Mobil.

The IRS has a different view of coin split-ups that occur when a blockchain forks into two chains. The split creates a windfall equal to the newly created coin's starting value at high ordinary-income rates.

When Bitcoin (BTC) spun off and forked Bitcoin cash (BCH) in 2017, the original BTC coin continued to live on one blockchain while the newly created Bitcoin cash, on a new chain, was dropped into the BTC owner's lap. You have to declare the value of BCH as ordinary income. It's a good bet that many coin holders neglected to do so.

How does the tax agency justify its rule? With some very strained logic. It sees a coin split as less like an oil company splitting in two than it is like a taxpayer stumbling on a $100 bill in a parking lot.

The new currency created by a fork is income when you can get your hands on it. Even if you hold on to the new currency, the new coins' cost basis is whatever you had to report as income.

Airdrops Create Ordinary Income

An "airdrop" is the random distribution of coins in the course of a marketing effort. (The IRS has also used the term, incorrectly, to describe the spin-off explained in the previous section.) With considerably more justification than taxing forks, the IRS considers marketing giveaways to be ordinary income.

You report the income from a marketing scheme as soon as you get the freebie. That reported income becomes the cost basis if you later dispose of the coins. The dollar amount will probably be small; people don't give away valuable coins.

Mining Creates Ordinary Income

Suppose you join a mining pool, spend $8,000 on electricity, and get rewarded with a Bitcoin worth $9,800. Even if you don't sell the coin, you have to report a $1,800 profit, and that profit is ordinary income.

Your new possession has a basis of $9,800, and any gain or loss from that point is a capital gain or loss. That could create a painful result. If the coin collapses in value to $8,000 and you sell it then, you have broken even, but you'll probably owe tax. That's because you'd be combining $1,800 of ordinary income, taxed at a high rate, with $1,800 of capital loss, which may be worth considerably less on your tax return.

The profit and loss described here apply if you are mining to make money. If, in contrast, the IRS can show that your mining is no more than a hobby, then you get stuck with hobby accounting. That's a disaster. Hobbyists must report all their revenue as income but can't deduct any of their costs.

Staking Gives Rise To Ordinary Income

Open-source platform Tezos rewards participants for putting up their coins as collateral and then certifying transactions known as baking. Bakers receive rewards of additional Tezos for validating transactions. Baking is the act of signing and publishing blocks to the Tezos blockchain. The reward coins are treated, like bank interest, as ordinary income. Some exchanges handle this work for you and then split the revenue. In that case, your income is your share of the fee, not the gross amount.

Gifts of Crypto to Charity

Buy a coin at $4,000, wait more than a year and donate when it's worth $9,000, and you get a $9,000 deduction without having to pay tax on the $5,000 gain. But gifts of property (as opposed to securities) worth more than $5,000 need appraisals, so this can get messy. If you donate appreciated property after holding it for less than a year, your deduction limits your cost basis. What about a depreciated property? Don't give it away. Sell it and take the capital loss.

Gifts Of Crypto To Friends

Cost basis and holding periods are the same as if you still held the coins, but with one small distinction: If the property has fallen in value during your ownership, then a special rule comes into play. Say you bought a Bitcoin at $12,000 and give it to your niece when it's worth $11,000. If she sells at more than $12,000, then she uses $12,000 as her basis. If she sells at less than $11,000, she has to use $11,000 as her basis, reducing the capital loss that she can claim. Any sale between $11,000 and $12,000 is in a dead zone that creates neither a gain nor a loss.

Tax Postponement Doesn't Work

In the like-kind rule, people aimed to treat one crypto exchange for another as a nontaxable event, postponing tax until the sale of the new coin. It probably didn't work for tax years before 2018 because coin exchanges didn't meet like-kind intermediaries' exacting requirements. It doesn't work for 2018 and beyond because a new statute limits like-kind treatment to real estate swaps.

Bitcoin Futures Are Section 1256 Contracts

Futures on Bitcoins, traded on the Chicago Mercantile Exchange, get the peculiar tax treatment of commodity futures: (a) Positions are "marked to market" on Dec. 31, with paper gains and losses recognized if the futures position was sold and immediately bought back. (b) The gains and
losses are assumed to be 60% long-term, 40% short-term, no matter how long the currency owner holds the position.

Crypto Is Probably Subject to the Straddle Rule

This rule forbids you to deduct a loss on closing a position in an actively traded investment (stock, option, whatever). In contrast, you maintain an open position that runs in the opposite direction. Thus, if you own an S&P 500 fund while simultaneously holding a short position in S&P futures, you can't sell just one of these to claim a capital loss while still holding the offsetting position. You could run into a problem with multiple positions in Bitcoin, Bitcoin futures, or Bitcoin options.

Offshore Crypto Not Subject To FBAR And FATCA Reporting

. . .

These two regulatory regimes compel you to disclose cash and securities held in offshore accounts. They don't, however, apply to property that isn't cash or securities. So your Bitcoin account at Malta-based Binance is not covered by these rules. Some lawyers advise you to file the reports anyway. If you trade during the year into conventional currencies (like dollars or euros), you might cross a threshold and be required to file. The labor cost of filing is small; the penalties for not complying are severe. The FBAR (Foreign Bank & Financial Accounts form), which kicks in if an offshore account tops $10,000 at any point during the year, must be filed electronically. The FATCA (Foreign Account Tax Compliance Act) has different thresholds that start at $50,000. The form, number 8938, can be filed on paper. You don't need to file these reports for assets held at a U.S.-regulated exchange like Coinbase. Exemption from account disclosure does not confer an exemption from the rule mandating the reporting of any sale again. If you have a profit from crypto, even a dollar, then it has to go on your tax return.

Identifying Lots Works As It Does With Securities

Say you buy five Bitcoins at $6,000 and five more at $8,000. Now you sell one coin for $9,000. No matter if it was one of the early ones (creating a $3,000 gain) or one of the late ones (a $2,000 gain)? The IRS gives you two choices. The default choice is first-in-first-out. In a rising market that tends to give you high tax bills. The second choice is "specific identification." You maintain meticulous records enabling you to spell out which coin sold — enabling you to make a selection that minimizes your tax bill (usually, the coin with the highest purchase price). It helps to have a coin tracking service to handle the dirty work.

CHAPTER SIX

HOW HAS BLOCKCHAIN LED THE DIGITAL REVOLUTION

In recent years, blockchain has been one of those technologies that have influenced our world socially, economically, and culturally. Along with Artificial Intelligence, Big Data, mobile devices, and social media. Blockchain will lead the digital revolution and introduces us to two main things: asymmetric cryptography and distributed systems that make blockchain so intuitive. When you understand asymmetric cryptography and distributed systems all the technical details of blockchain will make much more sense. Where blockchain is going is still hard to tell at this point but as the technology matures it undoubtedly is taking man from the Industrial Age into an age of revolutionary advancement we have yet to name. [1]

Blockchain provides us with the ability to create records that are indelible, transfer value by making updates, and automate these records through smart contracts among many other groundbreaking uses. This digital breakthrough creates a new and distributed platform that can give us the proper tools to transform our order of life and reshape the world of business. Many sectors of society from financial services, corporations, and governments to the music industry and phenomenon as the Internet of things, will be influenced by blockchain.

New technologies offer great potential in various sectors, but they must adopt the process in a balanced way. The blockchain is clearly a way of future development. The possibilities of applying this technology to other industries such as healthcare, retail, or real estate have opened great expectations about its use.

The blockchain is a mechanism, which is safe and transparent to carry out operations between two regions and even countries. In the bigger picture, it can assist governments in collecting taxes, deliver benefits, issue passports, and ensure the integrity of official records and services.

Basically, this technology is a step forward to eliminate the traditional way of storing records. Blockchain has the power to enhance entrepreneurship among developed and developing countries by removing the barriers built from embedded bureaucracy and corruption. Now, you have the ability to secure your online privacy by storing your digital footprints and also control access to it. It will be easier to fight crime and counterfeiting as every transaction can be independently verified and traced.

What Makes Blockchain a Vital Agent to Digital Transformation?

New technologies offer great potential in various sectors, but they must adopt the process in a balanced way. The blockchain is clearly a way of future development. The possibilities of applying this technology to other industries such as healthcare, retail, or real estate have opened great expectations about its use.

The blockchain is a mechanism, which is safe and transparent to carry out operations between two regions and even countries. Given the perspective of an even bigger picture, it can assist governments in collecting taxes, deliver benefits, issue passports, and ensure the integrity of official records and services.

Basically, this technology is a step forward to eliminate the traditional way of storing records. Blockchain has the power to enhance entrepreneurship among developed and developing countries by removing the barriers built from embedded bureaucracy and corruption. Now, you have the ability to secure your online privacy by storing your digital footprints and also control access to it. It will be easier and convenient to fight crime and counterfeiting as every transaction can be independently verified and traced.

U.S. Companies Are Investing in Blockchain

It's clear that many U.S. based financial companies are investigating how blockchain technology can streamline their current work processes and create more efficient networks to process financial transactions. Blockchain will not impact one or two industries its going to impact every industry. The same way the internal combustion engine changed every industry and propelled man into a new stage of advancement, much like the internet U.S. companies are getting an early footing in blockchain technology.

New companies like Bloq, and "old guard" companies like IBM, indicate a significant level of interest by U.S. companies in the technology. MicroStrategy the largest publicly traded business intelligence company revealed the purchase of 70,470 Bitcoins purchased at a price of $1.125 billion as of December,2020. That is a lot of money being pumped into the value and trust of the blockchain network. Many other big name corporations will follow in MicroStrategy footsteps as the financial benefits of blockchain are revealed more and more.

The sold-out Consensus in 2016: Making Blockchain Real conference in New York City was notable for its many big-name presenters from both inside (Gavin Andresen, Vitalik Buterin) and outside (Larry Summers, Delaware's Governor Jack Markell) the Bitcoin world. It was clear that the amount of "suits" at this conference signaled a major shift from Bitcoin and Blockchain conferences of old, which should not be construed as a bad thing. This pivot from an interest in Bitcoin to a new love affair with blockchain seems to be where American companies are focusing.

An Enabler of the Future

Blockchain is being embraced throughout the world by countries that are seeing its potential to not just disrupt existing financial systems but to solve existing financial concerns such as to provide banking for the unbanked, lowering transaction costs, and making cross-border transfers easier and more efficient.

In the U.S., companies are having a courtship and love affair with blockchain that can lead to solutions that either prop up their existing models or will create new innovative ways to address the current banking and financial system. If you're leaving that decision up to the companies that are part of the "status quo," you may end up with blockchain being little more than the "new database tool" rather the "next internet."

The Future of Contracts

Smart contracts are well suited for business activities that involve the purchase or exchange of goods, services, and rights, especially when frequent transactions occur among a network of parties, and counter parties perform manual tasks for each transaction.

This application matches many financial services transactions (e.g., simplifying automatic dividend payments, stock splits, and cryptographic signatures on stock certificates, streamlining over-the-counter agreements).

Despite various advantages and benefits of smart contracts and its use in the banking industry or the insurance industry, blockchain technology is still in its nascent stage. It will take time to go mainstream.

Before using/embracing the technology, the legal and regulatory aspects concerning the technology and smart contract needs to be looked into before it is recognized as a valid alternative to traditional contracts.

Smart contract platforms are intended to be more self-sufficient, self-governing, accurate, and transparent.

The benefits of taking one's business into the digital era are wide-reaching, and fraud prevention, reduced cost, and immutability of the smart contracts are unquestionably huge. Smart contracts can be used in all spheres of lives, from payment gateways to electricity bills, etc.

This practically guarantees that smart contracts will be a foundation of the future global economy and a part of every person's life. Here are some of the applications of smart contracts in everyday life.

Internet of Things Networks

There are areas where smart contracts intersect with other technologies, and the Internet of Things (IoT) is one of them. A combination of smart contracts and IoT is powerful and can enable significant transformations across industries, paving the way for new distributed applications.

Banking

Banking might be the primary industry where smart contracts appear to be the most significant alternative to the traditional transactions model. Smart contracts make payments as well as loans, and nearly all other financial operations are literally automated.

Insurance

. . .

The insurance industry can leverage blockchain technology in a significant way to transform their processes, such as claim fraud detection and claim settlement.

Real Estate

Automatically locking a house (through an Internet-enabled lock) upon a tenant's nonpayment of rent and then unlocking it when payment is submitted.

Agriculture

IoT sensors are reading the environment and automatically initiating activities such as irrigation or deployment of insecticide, based on programmed trigger values.

Supply Chain

Another area where smart contracts can provide real-time visibility is supply chains. Smart contracts ensure granular inventory tracking, benefitting supply chain financing, and reduce the risk of theft and fraud.

Legal Issues

The traditional model of resolving legal issues and certifying documents is also giving way to smart contracts. Smart contracts eliminate the need for notarization, offering not only an automated and unbiased but also a cost-efficient solution. *Io* illustrates the concept of notarizing documents using the Ethereum blockchain.

Health Care

Securing access to personal health records, enabling doctors to provide insurers proof of completed surgeries, supervising drugs and other supplies, and enabling secure and timely sharing of patient information for clinical trials and research.

Encoded and Decentralized Database

As mentioned, blockchain revolves around an encoded and decentralized or distributed database *(the 'distributed' part of distributed ledger technology)*, which serves as a ledger whereby records regarding transactions are stored, and cryptography is used for each update in transactions.

Blockchain technology is rooted in the world of cryptocurrencies, more specifically, Bitcoin. That connotation will disappear, and we will not speak about the blockchain but blockchains *(note the letter's')*, blockchain technology, or distributed ledger technology. Although today decentralization and the absence of a predefined central authority are often mentioned now in times where cryptocurrencies still have most attention, they aren't the strict essence *(moreover, even in Bitcoin, there are central authorities so to speak, the Bitcoin miners, but that's a different story)*.

Companies are urged to focus on private blockchains for commercial deployments *(Juniper Research)*

Blockchain technology is being tested and implemented across various applications, industries, and use cases for endless applications. Examples, on top of the Internet of Things and financial services *(banking, insurance, and reinsurance, capital markets)* include

Industry 4.0, fraud management, digital identities, information management, and far more areas and industries where it fits in a context of transactions, payments, contracts (smart contracts), proof, trust and so forth in the decentralizing nature of digital transformation technologies.

These records can't be changed as the model is distributed: there isn't a central authority, but they're also isn't any involved party *(those doing transactions)* that can change information. Blockchain relies on peer-to-peer network principles whereby each encrypted block in the chain is linked to the next. Why the peer-to-peer network and the absence of a central authority? Blockchain was precisely 'invented' to solve the lack of a central authority in cryptocurrency Bitcoin.

However, this doesn't mean that blockchain is only used for very decentralized applications. Blockchain is used by organizations and groups of organizations for specific services where trust from other parties is needed or to build a blockchain network with other parties without traditional intermediaries. It's also important to differentiate between public and private blockchains and between blockchain networks from the perspective of the relevance of the goal and context in which they are used, which organizations and groups of organizations use them, and what those exact services are.

The attention for blockchain from a security and secure transaction perspective is, among others, related to the fact that blockchain is a cryptographic ledger whereby the chain consists of encrypted blocks, and after the validation of the transaction *(peer-to-peer and across the network)*, it is added as a block to the chain as a permanent and unchangeable record of the transaction in digital ecosystems with heavy transaction processing whereby transactions, data, and speed increase and meet the need for a layer of trust.

---

1. Public-key cryptography, or asymmetric cryptography, is a cryptographic system that uses pairs of keys: public keys, which may be disseminated widely, and private keys, which are known only to the owner.
    A distributed system is a system whose components are located on different networked computers, which communicate and coordinate their actions by passing messages to one another.

CHAPTER SEVEN

WEIGHING THE PROS AND CONS OF BLOCKCHAIN

Many blockchain enthusiasts on the market are glorifying blockchain technology to be the next new disruptive digital technology. They aren't all wrong. In reality, this technology does change legacy systems, gets rid of intermediaries, and introduces a new world to us. However, many do fail to properly weigh the pros and cons of blockchain before they invest especially with a technology that is still evolving and growing.

The advent of every new technology, there will always come advantages and disadvantages; blockchain is on that list as well. So, when enterprises try to figure out blockchain's full potential, they often forget about the issues. As a result, many tend to make mistakes and lose resources, and waste time. So, to help you weigh the pros and cons of blockchain, I'll be covering them in this guide. So, let's see what the blockchain pros and cons are!

Pros of Blockchain Technology

. . .

There's a lot of pros to blockchain technology. Blockchain will change society in many domains. The potential to create trust on the internet with a real use case for record keeping anything is boundless. Check out the pros of blockchain as I break them down below.

Disintermediation

First of all, with blockchain, you get a distributed system. It means that it gets rid of any middlemen or intermediaries from your system. How is that an advantage? Intermediaries tend to be the third party source that connects you to your services.

However, in the current business world, every service the middlemen or intermediary offer, they get a cut or small fee percentage for conducting and merging your money, information or record to the source intended. That small fee payment might not seem like much for the individual, but it does add up when your service requires a 10-15 step process or your toggling hundreds of thousands of accounts..

Other than that, there's no way of knowing whether the middlemen would be honest with their services. In reality, corruption runs deep, and in many cases, these intermediates tend to abuse enterprises and consumers for their gain.

Thus, by getting rid of them, the whole trust issue is solved.

High-Quality Data

Blockchain technology offers a superior level of data quality. In reality, it is a distributed ledger system where it stores data. But how does it provide high-quality data? Well, you need to know that low-quality data would not convert into high-quality data within a night. That's not how it happens.

Anyhow, this distributed ledger technology offers a consensus process that allows you to filter out any bad data with useful data. It means that no one can add any information on the ledger or even manipulate the existing ones.

So, any new information is constantly being verified by the blockchain before getting added to the ledger, it will get rid of any false data if detected.

Moreover, it also gets rid of the issues that come from human-made errors. As information is verified, there's no scope for a human-made mistake until it causes quality control issues possibly costing large sums of revenue to correct.

Thus, blockchain increases the quality of data.

Let's move on to the next advantage in this blockchain pros and cons guide.

Durability and Security

Blockchain offers durability at its best. You could think of it as the internet where there's built-in robustness. In reality, the overall structure of the technology makes it so durable. As it stores blocks of information around the network, it makes sure that there no single point of failure or any single entity controlling it.

This quality makes the system inherently durable. More so, as no one can alter the blocks, it remains to be a solid secured platform. Other than that, it's quite efficient in fending off hacking attempts as well.

So, there's little to no possibility of overpowering this network.

Anyhow, let's see the next advantage in this blockchain pros and cons guide.

High Level of Integrity

Another great advantage of blockchain is the level of integrity. Compared to any other network systems out there, blockchain offers the highest level of integrity so far. But what does it mean? In reality, it means that all your data will always be the right one, and no one can alter them once it's on the ledger.

The process of storing the information and consensus processes are also robust. Any user can't just make changes to the verification as he/she pleases. Thus, it would offer accurate and reliable data every single time when you transact or store any other information. The hash ID of every blockchain plays a huge role in maintaining this property.

Immutability and Transparency

For the next advantage of blockchain pros, I'll be elaborating on transparency and immutability. Blockchain comes with an immutable storage system where you can't change any single form of data or let alone delete them completely.

In reality, cryptographic hashing plays a huge role in maintaining an immutable structure. As every single block will have a Hash ID, any changes to that block's data would change the ID drastically. And it's impossible to recreate the same Hash ID again.

Thus, if anyone tries to change the data, all the other users would notice right away. Think of the Hash ID as a link on a chain being pulled by two parties in a tug of war. If a link or HASH ID is cut, altered or replaced on any length of the chain all parties will feel the elastic recoil of the chain breaking transparently. Anyhow, most of the ledger system on this technology is also open for everyone to see. Even in private blockchains, there is common ledger information that anyone can see anytime.

Longevity and Reliability

. . .

As you already know, blockchain is immutable, transparent, and offers integrity. All of these characteristics result in the reliability and longevity of the technology.

Furthermore, as no one can change the blockchain's rules as they please, it remains intact. More so, as it can offer viable solutions for various business issues for the long term, it becomes a reliable technology.

Many enterprises are already considering altering their legacy networks with blockchain for the long term.

Simplistic Ecosystem

Many of you may find blockchain to be a complicated ecosystem; however, trust is a simplistic ecosystem. In reality, every enterprise needs to go through various stages to process or offer their consumers a solution.

What blockchain does here is to shrink down the various stages of processing into a few steps. More so, as the technology can offer almost everything online, it's much easier to maintain.

There are already many blockchain-based solutions on the market where they offer user-friendly interfaces.

Empowered Users

In the traditional centralized system, the users don't have that much control over their information. As a result, multiple corrupt personnel try to misuse the information for their gain.

However, as blockchain ensures a peer-to-peer network and gives users back the control, it ensures no one can misuse them in any way. The level of security and transparency makes sure that every individual is in total control of his/her information.

. . .

Faster Transactions

It also offers faster transactions compared to traditional means. Usually, the centralized banks can take a lot of time even to process a transaction. It became more eminent when someone tries to send money overseas.

It could take up to six days to even process that transaction. So, in times of an emergency, many consumers can't rely on traditional banks slow-paced system. However, with blockchain, you can complete a transaction within a few seconds!

That's faster than any traditional means so far.

Lower Transactional Costs

Besides offering faster transactions, it also offers a lower transactional cost—obviously, nothing is for free. When you use traditional methods to transact daily, you have to give them some form of fee for their services.

Even though the fee might be small, but after a multi-step process, it may become heavier in your pocket as the book stated earlier. On the other hand, blockchain only offers a lower transaction fee in exchange for a faster transaction process.

New Business Model and Value Chain

Another great benefit of blockchain in this blockchain advantages and disadvantages guide is the new business model. Blockchain comes with a new perspective on how we should model our business in the new world.

Thanks to the invention of blockchain, new marketplaces are being formed, and new opportunities are being created. Blockchain is a better infrastructure that gets rid of a lot of issues we face nowadays. It also adds more value to your enterprise once you start to use it by increasing revenues and promoting trust.

Improved Traceability

This is the last advantage in the blockchain advantages and disadvantages list so far. Well, it's more for enterprises that deal with the complex nature of the supply chain or any other similar industry—tracing back what or how you managed to produce an item is crucial.

In every stage, you need to ensure that it offers the highest quality. However, if you don't know your goods' origin, you can't offer quality. However, blockchain ensures that you can trace these items from the very source to the endpoint.

Not only good, but you can verify any asset in asset exchanges as well.

Blockchain Technology Cons

In these advantages and disadvantages of the blockchain technology guide, I'll now talk about the different blockchain technology cons. Let's see what these are.

Redundant Performance

Redundancy is the first con in the advantages and disadvantages of the blockchain list so far. In reality, the computation needs of this technology are more repetitive than centralized servers. It's because every time the ledger is updated, all the nodes need to update their version of the ledger.

It's because the distributed nature of the ledger system mandates that every node should have a copy of the ledger system. Thus, it needs to undergo the same process over and over again.

Complex Signature Verification Process

Another con in our advantages and disadvantages of the blockchain list is the signature verification process. Basically, for every transaction in the system, you'll need a private-public cryptographic signature verification.

It then uses the ECDSA (Elliptic Curve Digital Signature Algorithm) to ensure that the transaction happens between the correct nodes. Thus, every node needs to verify its authenticity, which can be a tricky and complex process.

Private Keys

To transact on the network, you will need to own a private key. Even though other users can see your public key, a private key is much more crucial as it remains hidden. Furthermore, all the blockchain addresses will have a private key.

You need to keep your private keys secured by any means if you don't want other people to misuse your assets. However, if you lose your private key, you will lose access to your funds on the network, as well. There's no way to recover them if lost or stolen. Keeping your private keys safe can be a disadvantage to users on the blockchain network if they are compromised.

Integration Concerns

. . .

Integration from old to new is also a major blockchain technology con. It's mainly for enterprises that run legacy networks. What are legacy networks? A legacy network is the generic name assigned to any old network not part of the TCP/IP protocol. Legacy networks are mostly proprietary to individual vendors. In reality, blockchain technology would replace legacy networks, most legacy networks are no longer used but many still exist. However, the integration process is still not fully functional. More so, many blockchain technologies can't work along with the legacy networks. It means that to use it properly, companies would have to get rid of their legacy networks for good completely.

Uncertain Regulations

Blockchain is new and still has large gaps of uncertain regulations. In reality, not all blockchain technologies come with the proper set of regulations on the network. Thus, many don't trust the system at all. On the other hand, the lack of regulation brings in the concept of ICO scams.

Needless to say, many have fallen victim to ICO scams as there is no regulation related to cryptocurrencies. Governmental institutions also struggle to adopt it as this sector runs fully on regulations.

Large Energy Consumption

To ensure that every transaction is valid, it needs to go through consensus processes. The consensus process requires a huge amount of effort to form every node. Not to mention, all the nodes need to communicate back and forth to ensure that a transaction is valid.

On the other hand, consensus algorithms such as proof of work require a lot of computational power, increasing overall power consumption. However, you'd be happy to know that now multiple consensus protocols consume much less energy.

No Control for Enterprises

Well, enterprises need a certain authoritative process to use blockchain technology. Unfortunately, public blockchains won't be able to offer these controlling aspects anytime soon. However, the rise of private and consortium blockchains seems to offer both the technology's control and distributed nature. So, we should be able to a more modern approach to this issue much sooner than expected.

Privacy Concerns

Another major con in the pros and cons of the blockchain list is privacy issues. Enterprises need their privacy to maintain its brand value at all costs. They can't reveal their sensitive information to the mass people or their competitors.

And so, many enterprises aren't that keen to use Blockchain for the enterprise. However, the use of private transaction options and user-specific authentication is solving this problem bit by bit.

Cultural Adoption/Disruption

Culturally blockchain will be the great disrupter shaking up many industries across the globe. Our current business models have been running on a specific architecture for quite some time now. But the invention of blockchain threatens that system.

In reality, blockchain already started to change how the system works and has disrupted many industries. As a result, multiple marketplaces have already become obsolete.

Lack of In House Capabilities

As crypto technology is a relatively new concept, there aren't many capable developers working on it. When developers working for fintech companies try to develop their enterprise blockchain solutions, finding a capable team to handle the project becomes difficult.

However, the rise of blockchain as a service can help you in blockchain business strategy. These services offer high-end developers and marketing teams to help you get your blockchain solution on the market.

High Cost

Yes, blockchain is much cheaper than other infrastructures. However, it can be a costly solution, as well. The cost depends on what type of feature you want to add and your needs. More so, developing a solution from scratch requires a hefty amount of money.

Not to mention replacing the legacy system would also cost a lot of money. However, you can overcome this issue if you keep your solutions at a minimum. Moreover, joining a consortium or using a BAAS can help you out as well.

According to a recent 2019 survey on 1386 global enterprises, it may surprise you what their adoption barrier can be. However, you should know that the survey does not include all the enterprises in the world that are working on blockchain technology.

Thus, the "true" percentage will always vary. If you want to get more information, you can check out the Deloitte survey form here.

Blockchain is a relatively new technology right now with a long journey ahead. So, it is quite eminent that it will have cons and along with pros as well. I've highlighted almost all the pros and cons of blockchain in this guide.

However, you should note that blockchain has already resolved most of the issues and figuring out a newer way to minimize the issues as much as possible over time. So, give it a couple of years more, and you might see a perfect technology ready to disrupt the industries.

CHAPTER EIGHT

AltCoins: How Can We Use Them?

In the mid-2010s, there was an explosion of altcoins flooding the market, and many shot up in price. It was resulting in many bearish investors stating volatility diluted Bitcoin's hype.

Today though, the hype around altcoins has mostly died down, and prices fluctuate less. Jesse Powell, the CEO of cryptocurrency exchange Kraken, believes that almost all altcoins are down 95% in value except during bull runs. Others speculate that there may be a cryptocurrency bubble that might burst in the future.

In this article, we'll look at what altcoins are and how you can profit by trading them.

What is an Altcoin?

The term altcoin comes from combining the words *"alternative"* and "*Bitcoin.*" Sometimes, crypto investors can refer to altcoins as "*coins.*"

At the time of writing, there are approximately 2,355 different cryptocurrencies. You can view them all on Etoro.

Bitcoin is not an altcoin because it dominates up to 66.3% of the cryptocurrency market, according to Coin Market Cap. So, with that in mind, all cryptocurrencies are altcoins, except one.

If an altcoin overtook Bitcoin, it would be interesting to see if the terms would change.

Ethereum and Ripple are the most successful altcoins.

The majority of altcoins will not survive; they will eventually fail and disappear.

While generally speaking, Bitcoin, Ethereum, and Ripple have occupied the 1st, 2nd, and 3rd largest coins by market cap, respectively, from fourth place onwards, cryptocurrencies have changed dramatically.

A coin that was fourth place a year ago or even a month ago could be replaced at a moment's notice.

Bitcoin is often considered the first generation cryptocurrency, and the altcoins that followed after are the second generation, such as Ethereum. Some cryptocurrencies, such as Cardano, are considered the third generation.

How Do Altcoins Work?

Generally speaking, altcoins work much like the original Bitcoin. Using a private key, you can send a payment from your digital wallet to another user's wallet. In a cryptocurrency such as these, there is a blockchain or recording ledger. The transactions remain permanently and publicly recorded, so exchanges cannot alter or deny after the fact —the blockchain mathematics secure proofs that confirm transactions in blocks.

Why do people keep creating new cryptocurrencies?

. . .

The primary goal behind many altcoins is Bitcoin's, to replace the current banking system in place and remove third parties from transactions.

People keep creating new cryptocurrencies because they believe that they can do better than Bitcoin.

Many altcoins are trying to solve Bitcoin's and Ethereum's problems, such as scalability and speed.

They do this by experimenting with different approaches to coin creation, fulfilling transactions and security, etc.

In some senses, we are still in the development stage in cryptocurrency. No one knows which method will be king.

By trading altcoins, traders are essentially betting on specific methods as being better than others.

Some altcoins like Tezos are well aware of this and have been created in such a way to ensure consumers can continuously update them.

Hard fork altcoins

Many altcoins have come to fruition by hard forking Bitcoin. A hard fork is essentially a split from another cryptocurrency, which creates a new one—for example, Bitcoin Cash. The first altcoin was Namecoin, running on the same code as Bitcoin in 2011.

Much of how Namecoin works is the same as Bitcoin, both using a proof-of-work algorithm and are both limited to 21 million coins.

Litecoin is another example of an early altcoin based on Bitcoin's code. It is an example of a cryptocurrency trying to improve on Bitcoin's lack of speed problems.

Many altcoins are ERC-20 tokens, which are cryptocurrencies built on top of the Ethereum blockchain.

Some altcoins that started as ERC-20 tokens have moved to their Blockchain.

. . .

Different types of altcoins

Altcoins are sometimes projects from enthusiasts and sometimes the basis for whole new businesses. They can even be more than coins, developing into entire new frameworks for everything from messaging applications to online marketplaces.

An altcoin will often change Bitcoin's rules sufficiently to do something uniquely productive and may have a particular application.

Some coins, such as solar coin, have been designed as a unit of exchange for solar power production. Others, such as name coin, have formed a new system of domain names on the Internet.

Stablecoins

Stable coins are a type of altcoin designed to combat cryptocurrency volatility by tying their value to an underlying index, commodity, or security. Tether is one example of a stable coin; Libra is a stable coin under development by Facebook.

Digital tokens

Altcoins that function as an underlying blockchain platform supports digital tokens. For example, Tether can also be considered a digital token built on Ethereum and other blockchains.

Some investors seek to earn returns by exchanging altcoins with each other, too, but it is risky as an investment. Virtual currencies trade on unregulated exchanges, which leave you vulnerable to price manipulation, fraud, and other problems.

Stablecoins are not particularly useful to traders to profit with because they peg to a reserve currency. Traders usually use them as a store of value. For example, after trading another altcoin, you may transfer the amount over to a stable coin to protect what you have made when you stop trading.

Stablecoins are popular with exchanges as a way to hold value. For example, many stable coins typically are tied to the US dollar.

Some companies will also refer to altcoins as "utility tokens" that corporations can use investor's capital additionally as a voucher.

Two examples of this would be LEO, used on the Bitfinex exchange, Binance Coin, and Binance exchange. Both coins give users a discount when they use them to make trades.

How can I get hold of altcoins?

Many exchanges do not allow you to purchase altcoins with fiat. Usually, you will have to do it through another cryptocurrency such as Bitcoin or Ethereum or a stable coin such as Tether.

Trading altcoins can get complicated because they may require specific wallets to store them. Some altcoins are only available to be held in the wallets created by their developers.

You may think that you can get around this by trading altcoins as CFDs via a broker; however, many brokers do not list little-known cryptocurrencies.

Even the best CFD brokers will likely not offer more than ten different cryptocurrencies.

Why do people trade altcoins?

Trading altcoins is a great way to diversify your portfolio. When one coin is down, you can trade another.

You can also hedge your trades by trading two cryptocurrencies that are very different.

The problem with altcoins is that you cannot use them in many places. Most businesses that accept cryptocurrency will usually only utilize Bitcoin. Because of this, the main reason people are interested in cryptocurrency is to trade it.

What should I look for when trading altcoins?

There is plenty of things to look for when trading altcoins. Liquidity is the key thing to look for when trading altcoins. You don't want to exchange an altcoin that doesn't move much at price because it will be challenging to get at the prices you want. If you decide to get involved with an altcoin early on in its development, it will be harder to trade. It may turn out to be more of an investment, as you may have to wait until it is valued enough to sell it. However, getting involved in a new altcoin early, though it can be very profitable, is very risky.

ICOs

You may decide to get involved in an altcoin during an ICO.

While doing your research, some good advice is checking out how well the ICO performs this. There are several sites online that rank ICO performance worth checking out.

Getting involved during the ICO stage is an excellent way to profit from altcoins. Typically, after an ICO, the price shoots up, and traders sell to make a quick profit.

Usually, these traders will not know much about the altcoin or care about the vision; they need to know that the price will shoot up.

That said, it is still highly advised that you know a good deal about anything altcoin before taking part in an ICO.

Today, many ICOs are held in private and offered to only large institutional investors, and we hear less about them than a few years ago.

That said, they still manage to move a large amount of money.

. . .

Altcoin vs. Bitcoin

Altcoins don't all follow the same rules as Bitcoin. For example, while Bitcoin will only ever mine, or produce, Bitcoins every 10 minutes, an altcoin called Litecoin will create coins every 2.5 minutes. This speed in coin production allows Litecoin to process payments faster. Litecoin will also produce 84 million Litecoins, whereas Bitcoin will only produce 21 million Bitcoins.

Litecoin also uses a different set of rules for mining than Bitcoin. Whereas Bitcoins require costly hardware to mine, anyone can mine Litecoin with standard computer hardware.

Litecoin is just one of the thousands of altcoins on the market. Some altcoins stand out as popular alternatives to Bitcoin, although they don't reach Bitcoin's $100 billion market cap.

Top five Altcoin Cryptocurrencies

Ethereum

The best altcoins list must begin with Ethereum, which is the second most valuable cryptocurrency after Bitcoin. It has a current market capitalization of $64.35 billion! Ethereum was launched in July 2015 by the now-famous prodigy of the cryptocurrency world — Vitalik Buterin.

Historical Investment Trends

You can see a snapshot below; the price of Ethereum grew from $9 in January 2017 to $1,389 in January 2018, giving a return of over **17000%**. Its price almost doubled in price last month, from $396 at the beginning of April (2018) to $648 today.

. . .

How is Ethereum unique and different from Bitcoin?

Bitcoin is just a digital currency, but Ethereum has more to offer in addition to that. Ethereum provides a platform to the developer on which they can build blockchain-based smart contracts and decentralized apps.

Ethereum was the first cryptocurrency to introduce smart contracts, which many consider as the next big thing.

Note: Smart contracts are agreements coded on the Blockchain that execute when certain pre-set conditions clear.

Ethereum is also the most preferred platform for launching Initial Coin Offerings (ICOs). For more on ICOs, see our *What is an ICO* guide.

These are some of the key things that set Ethereum apart from Bitcoin and attract both developers and investors.

Upcoming Events

One of the biggest challenges for Ethereum has been scalability. Ethereum's network currently supports roughly **15 transactions per second**. Transaction speed in the teens is not that great compared to VISA's 56,000 transaction messages per second.

The developers expect Ethereum to upgrade its technology from *Proof-of-Work* to *Proof-of-Stake* in 2020 with a 2.0 version. It is increasing its scalability to a large extent and customer base.

Where to buy Ethereum

Being one of the most widely available cryptocurrencies, you can purchase Ethereum from any popular exchange. Many of these exchanges support credit/debit cards in addition to bank transfers. You can buy Ethereum from Coinbase, Coinmama, CEX.io, or Gemini.

Litecoin

Litecoin, one of the oldest altcoins, was created in 2011 by an ex-Google employee, Charlie Lee. Like Bitcoin, Litecoin is also just digital currency but with improvements. The reason for launching Litecoin was to overcome some of Bitcoin's shortcomings, especially its slow transaction speed.

Historical Investment Trends

Unpredictability in the market causes coins to go in and out of the market quickly. Litecoin has been in the market for over seven years. It has grown to become one of the best altcoins with a market capitalization of $8.2 billion.

It gave a tremendous return in 2017 when it grew from $4 in January 2017 to $350 in December 2017. That's a return of 8000%!

How is Litecoin unique and different from Bitcoin?

Litecoin has made considerable improvements in Bitcoin's technology to increase the speed of transactions. The reason that it is called *Lite*coin is that it is four times faster than Bitcoin.

A Bitcoin transaction takes about 10 minutes to complete, but a Litecoin transaction completes in 2.5 minutes. Speed is making Litecoin more scalable than Bitcoin as it can achieve more transactions per second.

Litecoin has been built on Bitcoin's technology itself and is considered as the closest rival of Bitcoin, more so because both of them serve the same purpose: offering an alternative to fiat currencies like USD, EUR, CAD, etc.

Upcoming Events

. . .

Litepal, which is the first payment processor for Litecoin, got launched in February 2018. Buying, sending, and receiving Litecoin (or any cryptocurrency) is a very complicated process. LitePal payment processor will ease transacting in Litecoin currency much simpler. LitePal also provides services and support for merchants with easy integration to most existing eCommerce platforms.

As you can see in the picture below, this service has the potential to make Litecoin more mainstream and one of the most promising altcoins.

Where to buy Litecoin

Like Ethereum, it is relatively easy to purchase Litecoin due to its availability on most of the top cryptocurrency exchanges. You can buy Litecoin from Kraken, Coinbase, Coinmama, or Bitsquare.

NEO

NEO, which was initially called AntShares, was created in 2014 by Da Hongfei in China. It is the biggest cryptocurrency that has emerged from China.

Like Ethereum, NEO is a platform designed for developing Decentralized applications (Dapps), Smart contracts, and ICOs. Because of this close resemblance to Ethereum, NEO has earned the nickname the "Chinese Ethereum."

Historical Investment Trends

NEO has been one of the best altcoins of 2017 and has given tremendous returns. It grew from about $0.16 in January 2017 to about $162 in January 2018. That's about a 111,400% return on investment, which is enormous!

NEO is currently performing well as it has grown from about $44 at the beginning of April 2018 to $82 today. It has almost doubled!

One of NEO's significant factors driving growth is the Chinese government's support and its robust technology.

How is NEO unique and different from Bitcoin?

Like Ethereum, NEO also offers a platform for the development of smart contracts and Dapps. On the other hand, Bitcoin is just a digital currency that offers no such outlet.

NEO offers many technological advantages over Ethereum. NEO can handle about 10,000 transactions per second, whereas the Ethereum blockchain currently supports around 15 transactions per second.

NEO supports programming in multiple languages like C++, C#, Go, and Java, whereas Ethereum only supports one language — Solidity. NEO is quite popular among the developer community, as they can use NEO's platform in the language they already know rather than have to learn a new one.

Upcoming Events

In 2018, NEO's goal involved building an infrastructure to achieve its vision of building — *Smart Economic*. The smart economy's basic idea revolves around digitizing real-world physical assets like cars, houses, and much more.

These digitized assets can then be sold, traded, and leveraged through smart contracts. With such aggressive plans, a good team, and rising prices, NEO is one of the best altcoins 2020.

Where to buy NEO

. . .

There are not many options to directly buy NEO using fiat currency like USD or GBP. The easiest way is to buy Bitcoin from Kraken, Coinbase, and then exchange it for NEO. You can trade Bitcoin to NEO on many large exchanges like Bitfinex or Binance.

Cardano (ADA)

Cardano was founded in September 2017 by Charles Hoskinson, one of the co-founders of Ethereum. Cardano not only offers a platform for Dapps and smart contracts but also offers many technological improvements over Ethereum and other blockchains.

Historical Investment Trends

Cardano is one of the newest cryptocurrencies, but it has still managed to become one of the top altcoins with an $8.86 billion market capitalization.

While it has not given as high returns as other top coins, it is still popular among investors and developers because of its promise of building a highly robust blockchain that offers advantages over Ethereum.

Cardano grew from about $0.20 at its launch in October 2017 to touch a price of $1.20 in January 2018. That's a return of about 500% in 3 months.

How is Cardano unique and different from Bitcoin?

What makes Cardanounique is that it offers some improvements technology-wise over Ethereum and Bitcoin. Cryptocurrencies like Bitcoin are just digital currencies. Cardano provides a complete D-app-building platform while issuing its native currency called ADA.

The biggest problem of the 1st and 2nd generation small blockchain projects is that they have to struggle for improved scalability, interoperability, and sustainability.

Cardano has that covered to resolve international payment transfers, something that other low-end cryptocurrencies struggle to do. Cardano cuts down the transaction time of international payments from several days to a few seconds.

Where to buy Cardano?

Buying Cardano directly through a credit/debit card is not easy; there aren't many options. However, you can use Coinmama for this purpose. If that's not possible, you'll first have to buy Bitcoin or any other cryptocurrency and then use it to exchange for Cardno. Given that Bitcoin is more valuable, exchanging it unnecessarily is no good unless you need it specifically for building DApps.

EOS

Compared to other altcoins, EOS is relatively new. Developed by Dan Larimer, EOS's Initial Coin Offering (ICO) launched in June 2017. EOS was also responsible for creating the blockchain-based blogging site Steemit and the cryptocurrency exchange Bitshares.

Just as Ethereum, EOS gives its users a good platform for building decentralized applications. However, the technology it uses is different from what Ethereum offers.

Historical Investment Trends

In the beginning, EOS managed to raise a whopping $700 million through its ICO, making it one of the most ICOs in 2017. Even three years later, EOS's platform is still developing, yet that hasn't stopped the currency from dominating the alt-coins market.

Currently, the currency's market capitalization is nearly $15.69 billion, making it the 5th most valuable cryptocurrency. In June 2017, it started at $4, and by June 2018, it had grown to $18. EOS is quickly becoming an excellent altcoin to invest in 2020 as it surpasses its previous peak.

What makes EOS unique and different from Bitcoin?

Bitcoin is a blockchain and digital currency itself but cannot create smart contracts. On the other hand, EOS allows us to make smart contracts and DApps and be digital currencies. EOS, in many ways, is similar to the Ethereum platform.

However, EOS gives more options to smart contracts; Ethereum's only compatible programming language is 'Solidity.' You can create smart contracts in EOS through several languages, including C++. This small difference means many developers don't have to learn a whole new language to develop smart contracts.

Moreover, EOS and Ethereum don't share the technology used to make these contracts. Ethereum uses Proof-of-Work (PoW), whereas ESO relies on a mechanism called Delegated Proof-of-Stake (DPoS).

Delegated Proof of Stake (DPoS) vs. Proof of Work(PoW) vs. Proof of Stake (PoS)

Until now, we have only discussed the "proof of work" as the primary method consensus. However, there are alternatives for creating consensus without mining for coins. Delegated Proof of Stake is just one of those consensus methods.

A DPoS-based blockchain like EOS counts the voting of stakeholders who outsource their work to third-parties. Meaning they can choose third-party delegates to vote and secure the network on their behalf.

The third parties you trust become witnesses to transactions through your vote and help achieve consensus for block generation and validation. Your say in voting depends on how many coins you hold or how much 'stake' you have in the blockchain.

This unique consensus method stems from a broader consensus method called "Proof of Stake" (POS). PoS and DPoS use the same process for block creation/validation rights to stakeholders (your chances of being selected depend on the number of coins you own).

However, DPoS is different because it uses a system of elected delegates, whereas PoS primarily uses your stake in the Blockchain to assign block creation and validation rights.

That said, the voting system among participants varies from project to project. Just as elected national representatives each give individual 'proposals' or manifestos during their election campaigns, do these delegates when they ask for votes.

DPoS follows a democratic tradition, and each stakeholder gets to choose the delegate he or she trusts the most. The elected representatives usually share a portion of the rewards collected by their respective electors.

The DPoS algorithm devises this voting system based on a delegates' reputation. If elected nodes (delegates) don't work efficiently or misbehave, that node replaces one that works. This unique consensus technique can process transactions faster and more efficiently than PoW and PoS, making it more scalable.

Phew! That was a lot of information. But to summarize, the DPoS consensus system works well for platforms favoring PoS. Such systems can help make Blockchain more mainstream as people can use it to make their smart contracts.

Where to buy EOS

To buy EOS directly through US dollars, you can use Bitfinex. There are also multiple outlets where you can exchange Bitcoin for EOS. Similarly, the best use for EOS currency is efficient DApps and smart contracts.

. . .

The Pros and Cons of Altcoins

Advantages

Cryptocurrency veterans always see altcoins as contingencies to Bitcoin. They diversify your investment opportunities and give you something to fall back on if Bitcoin loses value. Moreover, most altcoins offer a unique function. The majority of altcoins have unique designs to help you create smart contracts and DApps.

These currencies have systems and process different from Bitcoin while simultaneously having greater scope to evolve. For instance, Po.et(POE) is an altcoin designed to give publishers and content creators a platform to easily license Ethereum and similar cryptocurrencies, such as amend flaws in Bitcoin. They are better in terms of speed, mining cost, and other factors. Altcoin XRP and the Ripple network have great technology, low energy consumption, and many nodes, which may become a significant competitor to other altcoins and Bitcoin in the future.

By changing blockchain rules, altcoin creators provide decent competition to the Bitcoin network, which is necessary to keep innovation going. Lastly, when used as a payment method, these cryptocurrencies offer even fewer transaction fees charged for each transaction while providing blockchain technology security features.

Cons

Although there are several advantages of altcoins, what usually holds them back is the lack of acceptance and exposure. Altcoins like Ethereum, Litecoin, and Bitcoin Cash do have substantial support, but they never match the scope of what Bitcoin offers.

Likewise, they face similar challenges in adoption, and they have only a limited number of outlets to support them. Being the largest currency and having the best support allows Bitcoin to overshadow most altcoins by a significant margin.

Since altcoins are mostly new, their value can be extremely volatile, too, in the realm of cryptocurrencies. Because of uncertainty in their value, many investors avoid investing too much. Moreover, investors are afraid of the high potential of scams and fraudulent schemes in altcoins.

Beware of Scams

Like any cryptocurrency in its nascent stage, altcoins are subject to skepticism from investors in the beginning. In many cases, creators of altcoins initially own a large amount of the circulating supply. Because of this, these altcoins can quickly become vulnerable to pump and dump schemes and insider trading.

Remaining extremely careful before you invest is crucial in any investment, especially the cryptocurrency market. Any promising cryptocurrency should have the vision to transform the financial system and make it transparent. If you see altcoins only promising to give you financial gains, they will be more likely to be scams.

Unless you find something extremely unique in a cryptocurrency, you should avoid trading it because it's probably not worth it. Suspicious altcoins won't feature smart contracts and try to repeat the formula of successful cryptocurrencies, telling you how they will do it.

Before investing in any altcoin, you should confirm that they have a blockchain and are decentralized. You can also consider altcoins that are open-source, and you can examine the code and how it functions.

Should You Invest in Altcoins?

Altcoins are widely popular with investors. Many of them favor altcoins because they consider them cheaper alternatives to Bitcoin that give a greater return. In reality, that doesn't happen often. Investors can always invest in smaller amounts of Bitcoin instead of buying an entire Bitcoin.

Likewise, historically, Bitcoin has been very kind to early investors and helped make a vast profit. Although some altcoins can earn you good profits, unfortunately, not all of them have the same potential.

New cryptocurrencies struggle to increase rapidly from $0.01 to something outstanding: $1,000, $100, or even $10 is considered a considerable gain. It's essential to do significant research before investing in any altcoin. Take heed to self-proclaimed gurus on Twitter and YouTube.

If people knew which low-cost altcoin would rise exponentially, they would never share it with other people and invest in it themselves. Moreover, never buy any cryptocurrency just after a sharp rise; instead, rely on your thorough research to make your investment decisions.

It's true that altcoins, in comparison, see higher profits than Bitcoin. However, most of the time, it happens only when the trend is rising. Whenever these trends start falling, altcoins get a hard hit. Investing in reliable altcoins and diversifying your investment portfolio can help secure investments, so making an informed purchase won't harm you.

CHAPTER NINE

Main Application Scenarios for Blockchain

Blockchain is a robust technology with plenty of applications in different domains. It goes far beyond just cryptocurrency and Bitcoin and can create a transparent and fair system while being more fast and efficient at the same time.

You can leverage blockchain in diverse sectors, from money transfer, payment, and asset management to contract enforcement and government transparency. In this part of the book, we will look into the main applications for blockchain closely.

Smart Contracts

We have discussed smart contracts in detail in this book, but not much on how real-world companies and people utilize them—smart contracts are just like regular contracts in the sense that they add levels of accountability for all parties.

However, the only difference is that the smart contract rules are enforced in real-time on a blockchain, thus eliminating intermediaries and increasing accountability levels for all stakeholders. These contracts save business time and money while ensuring compliance for all involved parties.

Contracts based on blockchain are increasingly becoming popular in public sectors such as government, healthcare, and the real estate industry. The following are the prominent examples of some of the most popular blockchain smart contract initiatives.

BurstIQ

Founded in Denver, Colorado, BurstIQ's is a significant data blockchain initiative centered on healthcare. It helps doctors and patients transfer confidential and personal information securely with each other. Both doctors and patients can establish parameters of data others can share. The contract even displays information on personalized health plans for every patient.

Media chain

Mediachain is an innovative smart contract initiative based in New York that helps musicians get the money they deserve through smart contracts. A decentralized contract is transparent, and artists can acquire more significant royalties and compel producers to pay the full amount on time.

The solution leveraged a decentralized, peer-to-peer database and connected applications with information and media. At the same time, an attribution engine generated cryptocurrency for creators to reward them for their work. The streaming giant Spotify managed to buy out Mediachain in April 2017.

. . .

Propy

Based in Palo Alto, California, Propy is a global real estate marketplace that has pioneered a decentralized title registry system for property owners. Property management can be strenuous unless you have a robust contractual system in place. The blockchain-backed online marketplace allows users to buy a property through cryptocurrency and makes title issuance instantaneous.

Money Transfer

The finance sector has primarily adopted blockchain since the rise of Bitcoin. After Bitcoin made it famous, cryptocurrency transfer apps are now becoming extremely popular. Financial companies are seeking blockchain solutions to save money and time, regardless of their size.

Money transfer applications allow companies to make real-time ledger systems, reduce third-party fees, and bypass the bureaucratic red tape. Computer World estimates that blockchain can help save the largest banks $8-$12 billion every year. Here are four of the top initiatives for money transfer blockchain companies.

OPSkins

Gamers are always into rare skins, gaming accessories, and emotes. It's a high-value market where some skins and accessories can cost thousands of dollars. OPSkins provides these gamers a reliable, transparent, and secure platform to exchange their favorite skins.

Users can even use Bitcoin as a method of payment in this online marketplace. Sellers can receive the Bitcoin they earn in their virtual wallets, and they have the option to either convert it into cash or keep the cryptocurrency. OPSkins manages to process as many as two million virtual transactions every week.

Circle

Circle internet financial is a Boston, Massachusetts blockchain-powered Fintech business that oversees more than $2 billion every month in exchanges between friends and cryptocurrency investments. The platform offers support for several currencies, including Bitcoin, Zcash, and Monero.

Chainalysis

Based in New York, Chainalysis is a cryptocurrency-cyber security tool that facilitates financial and governments to monitor the exchange of cryptocurrencies. Its advanced fraud detection system helps the company monitor transactions diligently and detect fraudulent trading, laundering, and violations in compliance. The tool ultimately helps to build trust in the blockchain.

Chain

Chain is a San Francisco, California based business that builds cloud blockchain infrastructures specifically for fintech companies. It uses cryptographic ledgers that safely and efficiently perform cryptocurrency transactions for financial institutions.

Internet of Things

. . .

Logically, the Internet of Things is the next step to blockchain's innovation. One of the main applications of blockchain is to erase security concerns. The Internet of Things (IoT) has several security and safety concerns. Increasing the number of IoT products gives hackers several endpoints to steal your data from every interconnected device, ranging from your smart thermostat to Amazon Alexa.

IoT infused with blockchain adds a greater security level and helps you prevent dangerous data breaches through the virtual incorruptibility of technology and blockchain's transparency. The following blockchain-powered companies are making IoT secure through blockchain:

Filament

Filament is a Reno, Nevada-based software and microchip hardware business that allows connected devices to run on the blockchain. They do this by encrypting ledger data and distributing real-time data to other blockchain-connected machines. The system runs by enabling the monetization of devices through timestamps.

Hypr

New York-based HYPR prevents cybersecurity risks in IoT devices by decentralizing credential solutions. Instead of storing passwords on a central server, the solution leverages biometric and password-free solutions, making its system virtually and the connected IoT devices virtually hack-proof.

Xage Security

. . .

Xage is one of the world's pioneers of blockchain-enabled cybersecurity solutions for IoT devices. The solution currently caters to billions of devices simultaneously and is advanced enough to self-diagnose and heal possible breaches. Most of the company's clients are IoT companies based in manufacturing, energy, and transportation.

Personal Identity Security

Life lock reported that as many as 16 million Americans reported identity theft and fraud in 2017 alone, which means users got their identities stolen once every two seconds. Fraud on such a grand scale results from hacking into personal files, forged documents, and several other ways to bypass security measures.

Securing sensitive information like birth certificates, social security numbers, and similar confidential information can help overcome these attacks. Decentralized blockchain ledgers allow you to do that and are effective, drastically reducing identity theft claims. Here are some of the top solutions that support identity security.

Illinois Blockchain Initiative

The Illinois Blockchain Initiative is state-funded and is currently in the experimental phase. The program has put several measures to enhance the security of social security numbers, birth certificates, voter registration cards, and death certificates using distributed blockchain ledgers.

Civic

. . .

Similar to several other security-driven blockchain initiatives, Civic focuses on information security. The civic wallet offers a solution to a unique ecosystem, giving people insight into who the public can access their information. Showing the customer's more control, access, and privacy over their data, including digital identity, health status, and Bitcoin quantity.

Users of this Civic enter into a smart contract with the company to individually decide who shares their personal information and to which extent. Users immediately get alerted whenever another user breaches the agreement in some way or an unauthorized source tries to access your private data.

Evernym

Rising from Salt Lake, Utah, is Evernym's Sovrin identity ecosystem. The blockchain-driven solution helps users manage personal information across the web through distributed web technology.

Sovrin allows all its users to store private information and behave as a communication medium between entities desiring confidential information while verifying correct information simultaneously at the same time.

Ocular

Ocular's anti-money laundering compliance platform leverages blockchain-enabled security to ensure hackers cannot manipulate data. The technology uses biometric systems to scan the faces of individuals applying for passports, driver's licenses, and other government-issued I.D.s. By viewing biometric systems on blockchains, governments can easily catch identity thieves foraging fake visas, certificates, and I.D.s from other countries.

. . .

Logistics

Logistic companies have massive networks stretched across continents. Moreover, a large number of shipping companies also make the logistics situation even more complicated.

A joint study by logistics leader DHL and Accenture discovered that there are more than 500,000 shipping companies in the U.S. alone. While these shipping delivery systems make every corner of the country more accessible, it also causes several transparency issues and data siloing. The same report went on to say that blockchain can help shipping companies resolve this problem and improve supply chain management.

This mammoth study proved extremely helpful in clearing up the situation and helping logistics and supply chain companies understand the importance of blockchain. Data transparency gives all participants of the supply chain a single source of truth.

Because all companies can share and acknowledge data sources, it's easier for blockchain to build greater trust within the industry. At the same time, blockchain can also open space for automation and make major processes leaner.

Although these efforts may seem unrelated to an outsider, the technology has the potential for saving logistics billions of dollars every year. Additionally, aside from offering security, blockchain provides cost-effective alternatives for the industry. The following are significant companies using blockchains to implement cutting-edge logistics solutions.

DHL

DHL is a world-wide logistics giant that offers shipping services to every corner of the world. Lately, the company has been experimenting with blockchain technology and is spearheading the blockchain revolution in the supply chain industry.

DHL's blockchain-based solution works through a digital ledger of shipments. This system maintains the integrity of transactions, keeping all shipping details transparent. Having a significant presence in the United States and the world, DHL has led shipping companies to embrace blockchain.

Block Array

Based in Chattanooga, Tennessee, Block Array has launched the "Bill of Lading" to operate its blockchain. It leverages a logistic platform for helping businesses monitor the progress of shipped orders and goods, manage payment details, and store information on drivers and materials. Lastly, the solution also offers secure document management and smart contract processing to secure all their orders.

Maersk

Maersk is another shipping giant based in Denmark and offers services across the United States. Maersk is making business moves to implement blockchain in its vast supply chain infrastructure; the solution teamed up with IBM (one of the world's largest tech companies) to infuse blockchain into global trade. The two companies will use blockchain to better understand the supply chain and track goods digitally across international borders in real-time.

Shipchain

ShipChain is a fully integrated blockchain system serving the end-to-end shipping process. From the moment the shipment leaves the facility to the time it arrives at its destination, the logistics ecosystem safely tracks and documents every move to create a transparent ledger. Based in Los Angeles, ShipChain aims to modernize the $8.1 trillion supply chain market using blockchain.

Government Use Cases

Surprisingly, blockchain has some of its strongest advocates in the government sector. The technology has tremendous potential for improving the government and ends the red tape culture efficiently. Governments such as the state of Illinois are already using blockchain for securing government documents.

However, blockchain's most lucrative use can eventually be in improving accountability, minimizing financial burdens, and creating efficiency in bureaucratic structures. Blockchain can potentially save government officials millions of hours every year through secure approval systems. The New York Times reports that blockchain can hold public officials accountable with smart contracts since it gives greater transparency and records all available records.

Likewise, blockchain can also revolutionize elections for the government. Turn-up at elections is not ideal. Only 58% of voters turned up for elections in the 2016 presidential election, whereas the 2014 midterm elections only saw 36.4% of its eligible voters turned up.

Creating a blockchain-based voting system can make the voting process more manageable and increase civic engagement since users can vote through mobile phones. Simultaneously, it will improve the transparency of the process, making the voting process more secure and incorruptible. Here are some of the significant companies leveraging blockchain technology for the government.

Voatz

. . .

Voatz is a Boston-based initiative that combines government, politics, and cybersecurity by creating a blockchain-based mobile voting platform. It includes encrypted biometric security, enabling voters to vote on a mobile device securely, regardless of where they are in the world. Fears of foreign powers influencing elections, a blockchain-based voting system eliminates the chances of data corruption and hacking. West Virginia first used this initiative, where the company collected from travelers and eligible service people during elections.

State of Delaware

Similar to the Illinois project, the state of Delaware has implemented its blockchain initiative. Centered in Dover, Delaware, the government program explores the benefits of blockchain in government and business.

Thus far, the program mostly focuses on storing confidential public documents to secure private records. In the next phase, the government plans to implement smart contracts between businesses and the government to improve transparency.

Follow My Vote

Secure voting platforms are not limited to Voatz. An open-source secure voting platform rising from Blacksburg, Virginia, has created a virtual open-source ballot box. Follow My Vote decreases the spending on physical ballots and makes voting accessible to every device. Since the project is open-source, it has a tremendous potential of blooming into a full-fledged system. With this solution, the government can give the voting process total safety and confidence.

Media Use Cases

. . .

Royalty payments, data privacy, and piracy are some of the minor issues in today's digital media. Deloitte reports that the widespread sharing of content after digitizing media has made it easier for others to commit intellectual property theft and infringe on copyrights. The report later shows us how blockchain can eventually help the industry ensure data rights and payments while mitigating piracy.

Blockchain can do this because it can prevent digital assets from existing in multiple places. In other words, it can help make digital media shareable while also ensuring that the media file preserves ownership. The transparent ledger system of blockchain is what makes piracy virtually impossible.

Moreover, blockchain can also maintain data integrity, allowing advertising agencies to tap into customers and musicians accurately to receive proper royalties for all original work. Here are some of the major companies currently investing in blockchain solutions for media.

Madhive

Originating from New York, MadHive is a blockchain-powered marketing and data solution specifically for digital marketing. Companies can save all customer data on private blockchains and use the solution to track and generate reports on customer activity. Combining analytics with secure blockchain enables you to perform real-time data monitoring and enjoy targeted audience reports without compromising data privacy.

Steem

Steem is unique in the line of blockchain-based solutions because it offers exclusive social media supported by blockchain. The system gives out "Proof-of-Brain" tokens as incentives and encourages all users to create original content for the community. Creators get tokens based on how many upvotes they receive for every article. Since June 2020, the organization has paid $60 million in tokens to creators.

Civil

The number of repressive governments has been increasing around the globe. Journalists face many hurdles while reporting events with transparency—civil aims to support journalists facing these challenges by empowering them through blockchain.

It uses CVL tokens to fund journalists and run independent newsrooms, freeing them from ad-driven editorial interferences. Through the Civil's model, journalists can operate in a decentralized system, enabling them to break critical news instead of relying on a central entity.

Open Music Initiative

Hailing from Boston, Massachusetts, Open Music Initiative aims to create an open-source protocol for identifying music rights holders and original creators. By trusting their music rights data to the blockchain, the nonprofit makes it easier for artists and musicians to be recognized for their work and paid correctly. The initiative has backing from virtually all areas of the music industry, including producers and radio stations and media giants like Netflix and Spotify.

Asset Management

. . .

The management of tangible, intangible, and complex assets is rapidly becoming more efficient with integrating blockchain technology.

Assets such as real estate - often constricted by liquidity and investment size - can be tokenized, divided, and distributed with minimal operational friction, management cost, and security concerns. Asset managers and novel investors alike can unlock opportunities with emerging liquidity and transfer dynamics.

Codefi

Asset issuance today is a slow and complicated process that involves multiple intermediaries. Legacy systems make institutions slow to react to ever-changing investor demands and increasingly stringent financial regulations.

Codefi streamlines the entire asset issuance and lifecycle management process. We offer rapid, secure, and customizable digital assets allocation, fully encoded, and automated with investor rights and obligations and compliance attributes.

The digitization of traditional securities and financial instruments in the future allow the monetization of a broader range of assets. Customizable issuance and rapid time-to-market allow issuers to match digital asset attributes directly with investor demands.

These advancements, compounded with fractionalized ownership and reduced operational costs, give rise to several game-changing benefits. Most prominently, it broadens access to a broader market of potential investors, increasing market size, and encouraging robust liquidity. It skyrockets secondary market opportunities, significantly reduces counter party risk, raises the funding process's velocity, and unlocks capital at lower costs.

CHAPTER TEN

Learning Powerful Crypto Trading and Investment Secrets: A Pro's Advice

Day trading cryptocurrency isn't for everyone, and there is a lot to consider before you get started. An estimate that almost 95% of all day traders eventually fail. That high percentage is why you must pay attention to crypto-trading investment secrets to get ahead.

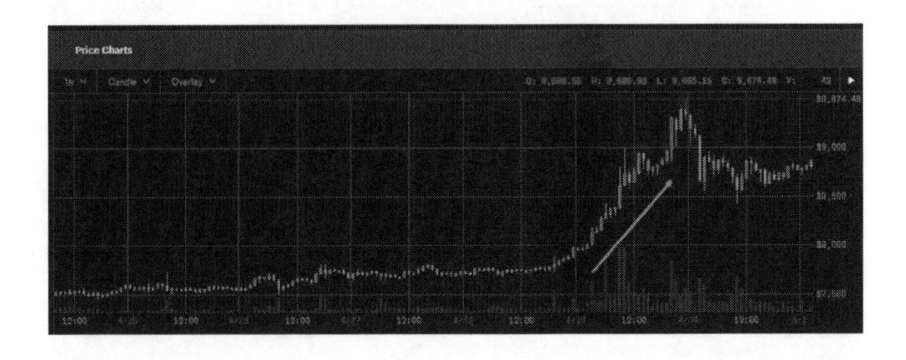

How to Read Crypto Price Charts?

The key to explaining a cryptocurrency chart's necessary details can be quite intimidating at first; let me tell you that the best way to learn, know, and understand a price chart is to try it. The best trading platforms offer demo accounts to all its potential and existing users to test any strategy and software. Let's talk about open, short, price fluctuations, and candlesticks. As you can see in the sections below, there are some critical components in a chart to look at.

Candles

Each bar you see in the chart represents a unit of the timeframe. If you are trading a 1-hour chart, each candle represents one hour, the same if you are watching a 15-minute frame, every candle will show you the performance in 15 minutes partitions. Red candles are drops in prices; green candles represent an increase in value.

Technical Indicators

Statistical studies are tools that can help your trading performance. You can add those studies to your chart by counting from the library of your platform.

Long Position

To open a long position, you buy the pair by licking in the buy box and choosing all the needed details. Then you speculate at a higher price.

. . .

Short Positions

When going short, you should click on the sell box and verify the information; you go short by believing the price will go down.

Candlesticks Charts

A chart builds with candles is called a candlestick chart. It is a visual tool that follows price fluctuations and is among the most popular type of charts in the cryptocurrency market.

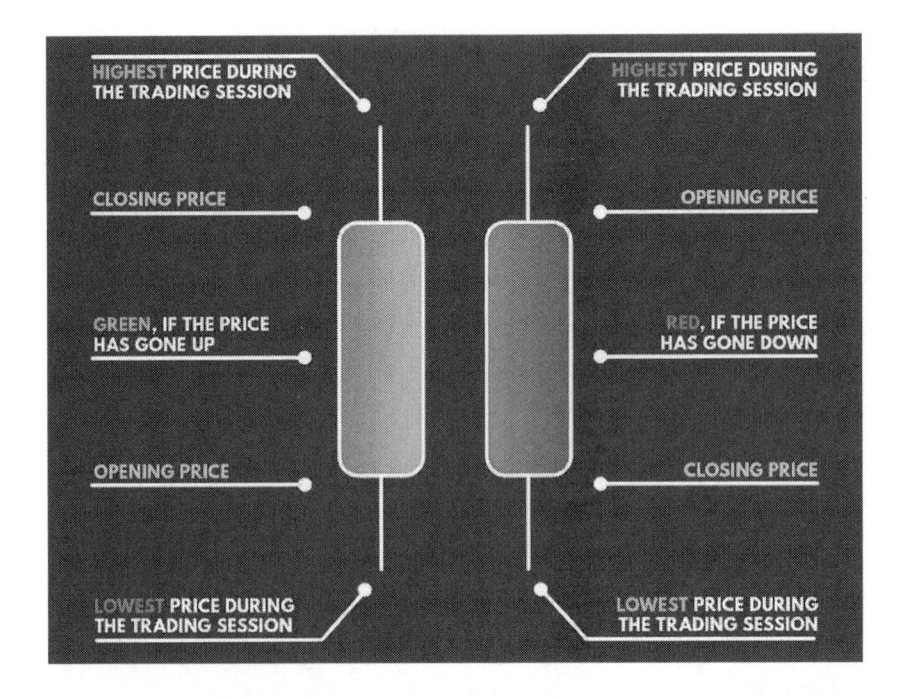

While a red candle is down, a green candle is when the price goes up. It has tails that represent the whole movement of price in that period.

The body shows the difference between the opening and closing prices.

What Affects Cryptocurrency Prices?

The prices of cryptocurrencies are affected by several factors; however, the most critical elements can make a digital asset fluctuate to the market's sentiment.

Regulation: All new crypto-related news and attempts to regulate existing and new digital assets can affect all cryptocurrencies fluctuations. For instance, every time advanced economies make laws about Bitcoin and altcoins, it will add pressure to the upside or downside to coins.

Forks: Forks or division and new tokens from an existing one is always a risky event for crypto traders. Usually, the fork's cryptocurrency will come under pressure in the days before the event and the hours after.

Technology updates: As cryptocurrencies continue to represent the innovative technology and the blockchain network behind it, any technology upgrade or implementation in the industry will push prices up.

Market sentiment: In the investment market, the view is everything; it is the base of every economy and the gas that fuels purchases and sales. It happens the same with cryptocurrencies. A good example is the 2017 rally, when everybody believes that it was the beginning of the crypto era, but then the big sale in 2018 when the very same people assumed it was too soon for a digital tokens economy.

. . .

What Are the Most Popular Cryptocurrency and Bitcoin Trading Strategies?

There are many different strategies to trade cryptocurrencies, but what is the best trading plan for everybody? As we told you before, there is no grail or secret sauce when investing, but hard work and hours in front of your charts.

There are three components for every operation of trading: Liquidity, volatility, and volume. If you pay enough attention, you will see it every moment you are in front of a chart. If you mix it well, you will find the right strategy for you.

Meanwhile, let's talk about three more popular techniques in cryptocurrencies.

Day Trading Cryptocurrency

A day trader is a person who opens both long and short positions and closes them within the same day. So, day trading is when your positions don't take overnight closes. You can use different timeframes, but the principal characteristic is that your trades don't live more than a few hours.

When you are doing day trading, you are watching for volatility and short term movements. It could be breakouts, scalping, or reversals. It also offers a higher return ratio, but at the same time, it involves the riskier bets.

Crypto Swing Trading

. . .

Swing trading is an investment technique that involves short to middle term investments. It works to make profits from extended movements or trends in cryptocurrencies.

As the trend is your friend, swing trading in cryptocurrencies goes from positions between two days and two weeks. Also, swing trades can maintain positions for months. The swing trader is always watching for pullback movements, resistance, and support levels.

Long-Term Trends In Cryptocurrencies

As the crypto market is relatively young, most of the movements are uncertain. The industry, brokers, and traders are still on the development of the nature of the market. For that reason, many experts don't trade cryptos in a long-term way.

However, if you are a believer in blockchain and cryptocurrencies, no matter the BTC/USD is at 1,000 or 10,000 per unit, you will see it as a cheap asset to buy. So, go ahead and invest in cryptos.

Long-term strategies in cryptocurrencies involve trends that develop in bigger timeframes such as daily, weekly, and even monthly charts. Swing trading is an excellent example of that; you take a direction and wait for its development until you see its exhaustion.

Then you can trade reversals or fading, which is when technical indicators such as volume, moving averages, and divergences show you that the trend is coming to an end and another is undergoing.

. . .

Best Tips Before You Start

Create a Game Plan

ow to me, this one seems obvious. It's too easy to say you have a plan; most of us have one loosely defined before entering a trade or purchasing a coin. But there's a world of difference between having a plan and sticking to it.

I know you've probably made the same mistake before too. You decide to change the plan at the last minute, and as a result, you sell too soon or too late.

You need to set criteria. Clearly defined rules you will abide by when the market goes one way or another. Every professional has a trading plan — and they follow it religiously.

So it's time for us to think like professionals. It's time to make a game plan. To help you get started, I've attached the template our team uses when making trades. It hits on all the key points the professional traders taught me.

After you've filled that in, I want you to print it and stick it wherever you do the most of your trading. It needs to be in plain sight, and when the market meets any of those thresholds, all you need to do is execute.

Having your strategy written down and in front of you makes it much harder to ignore.

It's also helpful to, on occasion, run through a sort of fire drill. Frequently question yourself, "If the market were to crash right now, and I lost everything. Would I be okay?" — if the answer is 'yes,' continue what you're doing. If the answer is 'no,' however, consider rethinking your approach.

Strengthen Your Portfolio

Go through your portfolio and separate the coins that have long-term potential from the coins that don't.

It's a bear market. The ship may not be going down just yet, but it's going through a rough patch. Any unnecessary weight is only going to drag you down. The market has no conscious and is very volatile; everything changes. You have real money at stake. If you don't move fast, you will lose it.

It is vital to act decisively when the market goes down. Consider what investments you have and be frugal. Put your money into a coin with better trading groups or into something you genuinely believe will withstand a crash.

Add this to your game plan document, list all your coins. List them in priority. When the market crashes, at what point will you begin to make cutbacks?

Ignore the Noise

You need to stay away from all the mumbo jumbo.

Long term profitability means positioning yourself either ahead or behind the crowd. Never in the crowd. Stay away from the chat rooms and the discussion boards. Everyone in these groups has an ulterior motive.

Far too many people spend their time discussing on feeds like Slack, Reddit, Telegram, Facebook, Discord, the list goes on. It would be best if you distanced yourself from these groups. There's too much misinformation floating around, causing you to make rash decisions and sell too soon.

Your only job as a trader is to watch the graphs and make predictions. It's not about gathering as much data as you can fit in your head to make accurate predictions. It's just about getting quality information; it doesn't have to be too much — it only has to be right.

For those of you who still want some informed data, some traders suggested joining VIP groups. These are close-knit communities looking to share insights and to make better trades as a collective.

Question the sources you follow. Ask yourself — is this necessary?

Am I following sources that aid my trading predictions? If not, cut them.

Discover Your Dream Team

Go out and find a few other people that trade, find people you trust. Start a group chat on Facebook messenger — or on whatever platform you prefer. Each of you should research 2-3 credible sources to begin following and then delegate the sources equally. Each day your team should study their respective origins and relay only the necessary information to the group, focusing only on information crucial for advising trading decisions.

Expect Volatility

Firstly, there is one significant difference between day trading cryptocurrency and day trading real-world assets. The reason for this is volatility. Volatility is when an asset's price moves up or down quickly, meaning it can either be a great success for the trader or a significant failure.

For example, if you were day trading stocks on the NYSE (New York Stock Exchange), it is improbable that the prices would change that much in 24 hours. They are safe companies that have been operating for a long time on the exchange. Of course, prices still go up or down, but it would generally only be by a small amount compared to cryptocurrencies unless you experienced economic depression, for example.

On the other hand, the prices of cryptocurrencies are very volatile. It is not unusual for a coin's price to rise or fall by more than 10%-50% in a single day. In some circumstances, even more. For example, in February 2018, a cryptocurrency called E-Coin increased in value by more than 4000% in just 24 hours, only to fall straight back down to where it started.

Buying the coin towards the start of the day would have made a lot of money; however, the people who bought it at its highest price would have lost most of their investment.

Avoiding Crypto Theft and Scams: What Experts Tell Us

A payment to third-parties as cryptocurrency transactions is irreversible. It is imperative to ensure the legitimacy of all involved third party merchants and third-party services before we send cryptocurrency to a blockchain address. You should only send cryptocurrency to entities you trust.

Just keep in mind, if anything sounds too good to be true, then it probably is. There is a high risk in transacting digital monetary mechanisms such as cryptocurrency. Cryptocurrency is still in the adoption phase. So volatility and the abundance of scams across the market, online, and cryptocurrency exchanges are an ongoing threat.

Types of Scams

Technical Support and Impersonation Scams

Fraudsters set up scam customer support and impersonate various companies—including Coinbase and Ledger—in the finance, tech, retail, telecom, and service industries. These scam phone numbers are spammed on the internet, luring unsuspecting victims seeking assistance. The scammers may also conduct outbound calls directly to potential victims. These scammers are skilled in social engineering, making false claims to deceive and manipulate their target to provide personal information used for fraudulent purposes.

• Never give support staff (or anyone else for that matter) remote access to your machine. Effectively provides the scammer with full access to your computer, online financial accounts, and digital life.

• Never give out your 2FA (2-Factor Authentication) security codes or passwords. Coinbase staff will never ask you to share sensitive authentication credentials.

• Never accept outbound calls asking for your confidential personal information. Be aware that scammers can spoof legitimate phone numbers when conducting outbound calls.

• Never send cryptocurrency to external addresses on behalf of alleged support agents. Coinbase staff will never ask you to send cryptocurrency to external addresses.

Giveaway Scams

Scammers are using social media to perpetuate giveaway scams. They post screenshots of forged messages from companies and executives promoting a giveaway with hyperlinks to fraudulent websites. Fake accounts will then respond to these posts, affirming the scam as legitimate. The fraudulent websites will then ask that you "verify" your address by sending cryptocurrency to the scam giveaway.

While it does offer a legitimate method for earning cryptocurrency, Coinbase does not engage in any giveaways. No one can reverse these payments, and Coinbase will never ask you to send cryptocurrency to external addresses.

• Never send cryptocurrency to giveaways under the guise of address verification.

• Be skeptical of all giveaways and offers found on social media. Do not trust screenshots in reply messages, as images are forged or altered.

• Use your favorite search engine to research any entity soliciting you on social media. If the offer sounds too good to be true, it probably is

. . .

Investment Scams

These scammers ask you to invest money to earn higher returns without financial risk and then request you to bring more people to do the same. They often need a constant flow of new people investing to make money. Ponzi and pyramid schemes are great examples of investment scams.

For the US, if you come across one of these scams, contact the Securities and Exchange Commission, the Federal Trade Commission, or your state's securities regulator to get help. For the UK, contact the Financial Conduct Authority.

Tips for avoiding investment scams:

• Be skeptical of websites or services promising high returns or unrealistic investment opportunities. If it sounds too good to be true, it usually is

• Only send cryptocurrency to trusted third parties. Search for publicly verifiable reviews or articles involving the recipient

• Watch for grammatical errors in communications or on websites. Scammers often make grammar or spelling mistakes.

• Research the organization thoroughly. Check consumer-protection websites and make telephone calls and send emails to verify authenticity.

Loader or Load-up Scams

Fraudsters frequently offer "loading" services on a variety of platforms. They claim to need Coinbase accounts with high limits, offering the victim a portion of the proceeds. These scammers use stolen credit cards on compromised accounts to perpetuate payment fraud. The result is the victim is left with payment delinquencies after the legitimate cardholder discovers the scam; the scammer often steals any available cryptocurrencies and submits unauthorized charges on veri-

fied payment methods. Be aware; you are responsible for any payments submitted using your authentication credentials.
- Never provide your passwords or security codes to third parties under any circumstances.
- Report any "loaders" to Coinbase and to the platform where they are advertising their credit card fraud

Telegram Scams

Coinbase has no official presence on the Telegram messaging platform. An abundance of scams on this platform target Coinbase users, including fraudulent payment bots and giveaway scams.

Employment Scams

Scammers will impersonate recruiters with fake job offers, actively seeking job hunters to steal cryptocurrency and personal information. The scammers will most frequently reach out to individuals who have posted their resumes online and ask for payment to begin training. These "job offers" often include convincing offer letters, and they may ask for confidential personal information.

Phishing

Phishing sites are malicious websites that mimic an authentic site to trick visitors into entering their login credentials or other sensitive information. These fraudulent websites distribute through various methods, including email, SMS text messages, social media, and search-engine advertisements.

. . .

Common Ways Scammers Entrap Cryptocurrency Traders

Here's a look at the more common scams and ways to avoid becoming a victim as you join the exciting future of cryptocurrency.

Imposter Websites

You may be following a solid tip from someone with a lot of expertise but still become a victim by accidentally visiting a fake website. There's a surprising number of websites that resemble original, valid startup companies. If there isn't a small lock icon indicating security near the URL bar and no "https" in the site address, think twice.

Even if the site looks identical to the one you think you're visiting, you may find yourself directed to another platform for payment. For example, you click on a link that looks like a legitimate site, but attackers have created a fake URL with a zero in it instead of a letter 'o'. To avoid this, carefully type the exact URL into your browser. Double-check it, too.

Fake Mobile Apps

Another common way scammers trick cryptocurrency investors is through fake apps available for download through Google Play and the Apple App Store. Although stakeholders can often quickly find these counterfeit apps and get them removed, that doesn't mean the apps aren't impacting many bottom lines. Thousands of people have already downloaded fake cryptocurrency apps, reports Bitcoin News.

While this is a greater risk for Android users, every investor should be aware of the possibility. Are there apparent misspellings in the copy or even the name of the app? Does the branding look inauthentic with strange coloring or an incorrect logo? Take note and reconsider downloading.

Bad Tweets and Other Social Media Updates

If you're following celebrities and executives on social media, you can't be sure that you're not following impostor accounts. The same applies to cryptocurrencies, where malicious, impersonating bots are rampant. Don't trust offers that come from Twitter or Facebook, especially if there seems to be an impossible result. Fake accounts are everywhere. Tesla CEO Elon Musk is an excellent example of this; due to Twitter hacks posing as Elon Musk's account, scammers stole hundreds of thousands of victims' Bitcoins.

If someone on these platforms asks for even a small amount of your cryptocurrency, likely, you can never get it back. Just because others are replying to the offer, don't assume they aren't bots, either. You have to be extra careful.

Scamming Emails

Even if it looks exactly like an email you received from a legitimate cryptocurrency company, take care before investing your digital currency. Is the email the same, and are the logo and branding identical? Can you verify that the email address is legitimately connected to the company? The ability to check on this is why it's crucial to choose a company with real people working for it. If you have doubts about an email, ask someone who works there. And never click on a link in a message to get to a site.

Scammers often announce fake ICOs, or initial coin offerings, as a way to steal substantial funds. Don't fall for these fake emails and website offers. Take your time to look over all the details.

. . .

As you start to invest increasingly at different exchange platforms and startups, you must be extremely careful not to lose your investments. Experts suggest confirming whether a company is blockchain-powered before you invest anything in cryptocurrency.

Meaning you must ensure that all transaction data is detailed. At the same time, it's essential to check your crypto investment has a solid business plan and that it solves real problems. They should specify whether their digital currency follows all ICO rules and estimate the currency's liquidity. You must ensure that there are real people behind the venture. Investing in a startup that lacks these characteristics is risky, and you must think through your decision hard.

Unfortunately, there are many ways that some Internet users exploit unsecured computing systems to mine or steal cryptocurrency. Learn more about staying safe and protecting yourself in this emerging market before you start investing in cryptocurrency.

How Can We Prevent Crypto Scams?

Investing in cryptocurrency earns you great rewards, but without the right precautions, you can lose your money in just a few clicks. You must ensure that your cryptocurrency exchange uses a blockchain-powered platform to process your transactions.

Blockchain-powered platforms use smart contracts and don't rely on human intervention, making digital transactions easier. The system's protocol reliably checks the legitimacy of transactions, making it harder to pull off frauds.

However, even then, fraud can happen to you. Our last section will look at ways to prevent scams and protect your funds from common hazards of crypto security.

Crypto Wallets

. . .

Experts suggest using crypto wallets to limit scams while trading cryptocurrency. Essentially, a crypto wallet is just a software program that helps us manage cryptocurrency. These wallets let you monitor their balance, store public and private keys, interact with various blockchains, and send and receive digital currencies.

Crypto wallets are necessary for preventing crypto scams and keeping your hard-earned money secure. However, they are available to us in two types; they are either hot or cold.

A crypto wallet connected to the internet and accessible ubiquitously is called a hot wallet. In contrast, a cold wallet allows you to store funds offline but can't connect to the internet. Such wallets will enable you to receive funds but can't transfer them out.

Another difference here is that all hardware wallets are cold wallets because they are offline. They usually come as USBs, paper wallets, and other data storage devices. Likewise, they are also available in the form of physical items such as physical Bitcoins. Hot wallets, on the other hand, are online. Therefore, they are mostly mobile, cloud, or available as software.

Traders and investors use both these types of wallets because they allow you to save cryptocurrency with versatility. Cold wallets will help you secure long-term crypto assets more effectively, whereas hot wallets are excellent for trading.

Looking for crypto wallets, you will discover four main categories: online, cloud, paper, and hardware. However, before discussing different types of wallets, you should remember to avoid saving your digital assets at cryptocurrency exchanges.

Cloud Wallets

Online wallets, by definition, are vulnerable. Your coins are stored in a cloud wallet, meaning a creative hacker can access them from any computer, device, or location. They are super convenient, but they hold your private keys online controlled by third-parties. Therefore,

they are more vulnerable to attacks and theft by design. Popular cloud wallets include:
- Guarda
- Coinbase
- Metamask
- Blockchain.info

A safer version of cloud wallets is **non-custodial online wallets**. They are accessible via web and apps, but the service provider does not have access to your private keys. In most cases, not custodial wallets are a part of the exchange platform, meaning that they let you trade your coins safely and securely. Examples of non-custodial cloud wallets include wallets by:
- Crypto.com Defi wallet
- LocalCryptos
- Bitwala

Software Wallets

Software wallets are downloaded and installed on a personal computer or smartphone. They are hot wallets. Both desktop and mobile wallets offer high security; however, they cannot protect you against hacks and viruses, so you should try your best to stay malware-free. As a rule, mobile wallets are way smaller and simpler than desktop wallets, but you can easily manage your funds using both of them. Besides, some software wallets allow you to access funds via multiple devices simultaneously, including smartphones, laptops, and even hardware wallets.
- Jaxx
- Freewallet
- Exodus
- Electrum Wallet
- Infino Wallet

. . .

Hardware Wallets

Unlike software wallets, hardware wallets store your private keys on an external device like a USB. They are entirely cold and secure. Also, they are capable of making online payments too. Some hardware wallets are compatible with web interfaces and support multiple currencies. Designed to make transactions straightforward and convenient, so all you need to do is plug it in any online device, unlock your wallet, send currency, and confirm a transaction. Hardware wallets are considered the safest means of storing crypto assets. The only drawback is that they aren't free to use.

Popular hardware wallets include devices by:
- Ledger
- Trezor
- KeepKey

Getting a hardware wallet directly from a manufacturer is the most secure way. It is unsafe to buy it from other people, especially the ones you don't know. Mind that even if you get a hardware wallet from a producer, you should always initialize and reset it yourself.

Typically, your wallet choice depends on your portfolio. Every cryptocurrency should have its native wallet, found on its website, but sometimes it may be more convenient to have a multi-currency wallet. Keep in mind that not all multi-currency wallets support all coins. Even hardware wallets have a limited amount of coins they support. On the other hand, there's no shortage of wallets for popular cryptocurrencies like Bitcoin or Ethereum.

Once you get a (hardware) cryptocurrency wallet, you will also need to protect your private recovery seed phrase. One of the most reliable seed word protection tools is CryptoTag, which allows you to store them on virtually indestructible titanium plates.

. . .

Paper Wallets

Paper wallets focus on cold storage. The term "paper wallet" generally refers to a physical copy or paper print of your public and private keys. It also means the software used to generate a pair of keys and a digital file for printing. Whichever the case, paper wallets can grant you a relatively high level of security. You can import your paper wallet into a software client or simply scan its QR code to move your funds.

If a paper wallet is available for the cryptocurrency of your choice, you're likely to find a guide on how to make one on the project's website or community page. MyEtherWallet is a universal way to make a paper wallet for Ethereum and all ERC-20 tokens. Use Bitcoin Paper Wallet Generator to generate a paper wallet for Bitcoin.

Although paper wallets are cold, they come with their share of risks, too. For instance, paper wallets can be easily damaged, burned, easy to copy, take pictures, and require mutual trust if you're not making one yourself. To make paper wallets less fragile, sometimes people laminate them, create multiple copies, store them in different locations, engrave them on pieces of metal or other sturdy materials, etc.

Note that it is a bad idea to keep electronic copies of your paper wallet on your PC. Private key owners of a paper wallet should always remain offline. Keeping your paper wallet files online makes it as secure as a hot wallet.

Best Security Practices to Secure Private Wallets

Securing your digital assets in a private wallet can significantly improve your chances of securing them. That said, doing so doesn't guarantee that it will remain safe. You must connect it with extreme care and ensure private keys remain private. Following the techniques below, you can ensure that your keys are secure.

Multi-factor Authentication

Multi-factor authentication is a reliable way to ensure only you can access your crypto account. It adds a security layer to your account and enables you to leverage separate channels to guarantee your account's exclusive access. Since crypto trading involves vast sums of money, leveraging hardware or software for multi-factor authentication is ideal.

Whitelist IP and Withdrawal Addresses

If you are using a static IP for additional security online, it's best to whitelist IP addresses for your account. You must ensure that only you access your funds and accounts.

Double-check Crypto Addresses

Some malicious programs can edit and paste a wrong transaction address whenever you send a transaction. Typically, the new address belongs to an attacker. It's better to be safe than sorry.

Use Security Measures You Can Handle

Some people never feel secure and go to the furthest lengths to secure their cryptocurrency. However, they forget that they can also lose crypto to their security tools. Losing access to your accounts, funds, or wallets is as common as hacks. Don't overcomplicate your security if that's not what you're into anyway. Strive for an appropriate balance between complexity and safety.

Stay Clear of Phishing Sites

Whether you're connecting to exchange or online wallet, confirm that you're logging in to the right address. Many bogus websites imitate exchanges for the sole purpose of stealing your login data. Always check whether the website address is correct.

HTTPS

Login only to secure websites with a valid HTTPS certificate. Most legit sites have one. For extra safety, try browser plugins like "HTTPS Everywhere."

Ensure That You Have a Secure WiFi

Never connect to your online wallet, exchange account, or another critical security point via public WiFi. Even when you're at a presumably safe place, make sure your WiFi access point uses strong encryption like WPA-2 protocol.

Don't Keep Digital Assets in One Place.

. . .

Don't keep all your crypto assets in one place. The best way to handle it is by using one or several cold storages for long-term holdings and at least one hot wallet for trading and transactions.

There are also several other precautions you must take to save your private keys from getting compromised:

- Never tell anyone about your crypt assets publicly using your real name or through an identifiable address. Doing so can endanger your assets from burglars who steal crypto funds in offline storage.
- Avoid storing cryptocurrency at exchanges for prolonged periods.
- Only trust your hardware wallet screen to validate information on the device.
- Always enable multi-factor authentication function.
- When using a hardware wallet, select a pin code that no one can guess easily and avoid putting the 24-word recovery code online.

When cybersecurity is one of the most significant concerns for businesses, safeguarding cryptocurrency, and securing digital assets is a critical skill we must master. Following these tips will help you securely trade cryptocurrency and limit the chances of fraud and security breaches.

ABOUT THE AUTHOR

Dawdu M. Amantanah, a best-selling author of several books Death of a Rose: Rise of the Black Petunia, Cover Me In Gold: The History, Horrors & Other Unknown Facts About Gold, and the forthcoming The Bit on Digital Coins: A quick practical guide to understanding cryptocurrency. An alumnus of Southern New Hampshire University School of Business before discovering a passion for writing, Dawdu worked for technology and financial institutions such as J.P. Morgan Chase and State Street Bank.

Made in the USA
Columbia, SC
23 January 2025